Analytical Biotechnology

A C S   S Y M P O S I U M   S E R I E S **434**

Analytical Biotechnology
Capillary Electrophoresis and Chromatography

Csaba Horváth, EDITOR
Yale University

John G. Nikelly, EDITOR
Philadelphia College of Pharmacy and Science

Developed from a symposium sponsored
by the Division of Analytical Chemistry
at the 196th National Meeting
of the American Chemical Society,
Los Angeles, California,
September 25–30, 1988

American Chemical Society, Washington, DC 1990

Library of Congress Cataloging-in-Publication Data

Analytical biotechnology: capillary electrophoresis and
chromatography
    Csaba Horvath, editor, John G. Nikelly, editor.
    p.   cm.—(ACS symposium series, ISSN 0097–6156; 434)

    "Developed from a symposium sponsored by the Division of
Analytical Chemistry at the 196th National Meeting of the
American Chemical Society, Los Angeles, California, September
25–30, 1988."

    Includes bibliographical references and index.

    ISBN 0–8412–1819–6

    1. Capillary electrophoresis—Congresses. 2. Chromatographic
analysis—Congresses. 3. Proteins—Analysis—Congresses.
4. Biotechnology—Technique—Congresses.

    I. Horvath, Csaba, 1930– . II. Nikelly, J.G. (John G.)
III. American Chemical Society. Division of Analytical
Chemistry. IV. American Chemical Society. Meeting (196th:
1988: Los Angeles, Calif.) V. Series.

TP248.25.C37A53   1990
660'.63—dc20   90–40212
                                             CIP

The paper used in this publication meets the minimum requirements of American
National Standard for Information Sciences—Permanence of Paper for Printed Library
Materials, ANSI Z39.48–1984.

PRINTED IN THE UNITED STATES OF AMERICA

Foreword

The ACS SYMPOSIUM SERIES was founded in 1974 to provide a medium for publishing symposia quickly in book form. The format of the Series parallels that of the continuing ADVANCES IN CHEMISTRY SERIES except that, in order to save time, the papers are not typeset but are reproduced as they are submitted by the authors in camera-ready form. Papers are reviewed under the supervision of the Editors with the assistance of the Series Advisory Board and are selected to maintain the integrity of the symposia; however, verbatim reproductions of previously published papers are not accepted. Both reviews and reports of research are acceptable, because symposia may embrace both types of presentation.

Contents

INDEXES

Preface

ADVANCES IN BIOTECHNOLOGY HAVE BROUGHT ABOUT novel and efficient means for industrial production of therapeutic proteins and other biological substances. As a corollary to these accomplishments, the need has arisen for rapid analytical techniques with high resolution and sensitivity to facilitate research and development, process monitoring, and quality control in biotechnology. Such applications have provided the major driving force for the rapid growth of new methodologies for biopolymer analysis in the past few years.

In fact, we have witnessed the birth of new analytical methods that play an essential role in biotechnology. Further progress in life sciences also depends on the introduction of increasingly sophisticated analytical tools for the separation, characterization, and quantitative assay of complex biological molecules. Consequently, the concepts, methods, and strategies of analytical biotechnology will inevitably be adopted by the life sciences.

In the manufacture of proteinaceous drugs, the purity of the final product is of paramount importance. Purity, however, is by no means an absolute term; it depends on the method used for its measurement. Since traditional chromatographic methods and slab gel electrophoresis have been the main tools for biopolymer analysis, the commonly used but utterly vague terms *chromatographically* or *electrophoretically* pure demonstrate the role of available techniques not only in the measurement but also in the definition of purity.

In the past decade, we have benefited from major improvements in the sensitivity and efficiency of both chromatography and electrophoresis. High-performance liquid chromatography (HPLC) has become firmly ensconced as a powerful method for protein analysis. Ironically, reversed-phase chromatography, which employs denaturing conditions, has found a particularly wide application, at least when only analytical information is sought. This is the case in most routine analytical work required for many biotechnological applications.

The advantages of HPLC over classical chromatographic methods stem from the employment of a precision instrument that utilizes high-performance columns with concomitantly high analytical speed and resolution and affords total control over the chromatographic process and sensitivity of analysis. In a way, the recent emergence of capillary electrophoresis (CE) follows the same patterns: electrophoresis, a well-established and widely used method of biopolymer analysis, is carried out by a suitable instrument that exhibits some of the major features of a liquid chromatograph. Although numerous technical problems are yet to be solved, it is not difficult to anticipate that instrumentation of electrophoresis will bring forth the advantages also germane to HPLC and thus will greatly expand the potential of electrophoretic analysis.

This volume accounts for some of the recent developments in CE and HPLC that are of particular interest in biotechnology. Four chapters dealing with capillary electrophoresis present an introduction to this technique and discuss its application to various analytical problems ranging from the analysis of cyclic nucleotides to quality control in the pharmaceutical industry. Another four chapters encompass recent developments in HPLC and has a sharper focus: analysis of pharmaceutical proteins. This particular area is of vital importance to biotechnology, and recent progress has been quite impressive. Enhancement of analytical performance in terms of speed, resolution, and sensitivity combined with the integration of various techniques has engendered powerful schemata for routine assay of proteins at purity standards unheard of a decade ago. An adjunct chapter elaborates problems associated with the use of HPLC as a detection method in preparative chromatography, an approach that will surely find increasing application in industrial protein purification. Mass spectrometry has found growing employment in the structure determination of peptides on a routine basis, and one chapter deals with the use of this technique. The last contribution in this volume is aimed at the use of the displacement mode of chromatography, which is primarily a preparative technique. Its potential in analytical work, however, is being more and more recognized for the enrichment of trace components.

We thank the authors for their contributions, which made possible the publication of this volume. Analytical biotechnology continues to face new challenges, and its armory of adequate tools for meeting them is far from being complete. The collection of papers presented here should be viewed as a testimony for both the incipient nature and the vast potential of this field, which does not yet warrant the publication of a treatise.

CSABA HORVÁTH
Yale University, Department of Chemical Engineering
New Haven, CT 06520

JOHN G. NIKELLY
Philadelphia College of Pharmacy and Science, Chemistry Department
Philadelphia, PA 19104

May 18, 1990

Chapter 1

High-Resolution Nanotechnique for Separation, Characterization, and Quantitation of Micro- and Macromolecules

Capillary Electrophoresis

Noberto A. Guzman[1,4], Luis Hernandez[2], and Shigeru Terabe[3]

[1]Protein Research Unit, Princeton Biochemicals, Inc., Princeton, NJ 08540
[2]Department of Physiology, School of Medicine, Los Andes University, Merida, Venezuela
[3]Department of Industrial Chemistry, Faculty of Engineering, Kyoto University, Kyoto, Japan

A powerful high-efficiency, high-resolution analytical technique is described for the separation, characterization and quantitation of minute amounts of analytes. This technique, termed capillary electrophoresis, offers the capability of on-line detection, the use of multiple detectors, micropreparative operation and automation.

The determination of minute quantities of micro- and macromolecules is an important problem in biological chemistry and poses a challenge to biological chemists. Attempts to optimize separation and characterization conditions and techniques have always been a major concern to many scientists. Unfortunately, most of the advanced new technologies currently available to biological chemists still require microliter quantities and hardly reach subpicomole sensitivities.

Two of the most powerful separation techniques used today are chromatography and electrophoresis. Although various modes

[4]Current address: Roche Diagnostic Systems, Inc., 340 Kingsland Street, Nutley, NJ 07110–1199

0097–6156/90/0434–0001$09.75/0

of chromatography are used for separation and characterization of macromolecules, quite often the final purity test is performed through electrophoretic analysis. If a single peak is obtained during the chromatographic analysis of proteins and peptides, electrophoresis will probably be used as a confirmatory purity test. The opposite is unusual.

Although the separation modes of the electrophoretic methods practiced today are many (1,2), they are slow, labor-intense, prone to relatively poor reproducibility and have limited quantitative capability. In addition, it has been difficult to accomplish a fully automated operation. On the other hand, the emergence of capillary electrophoresis (CE) gradually has begun to solve problems in which the handling of low nanoliter samples and subfemtomole quantities is necessary. Furthermore, among the major advantages of capillary electrophoresis is that it can be made fully automated, it has high resolution capability, and it can quantitate fully minute amounts of sample to be analyzed. Because a significant amount of information has been reported during the last decade about capillary electrophoresis (for recent reviews see 3-9), we are aiming to update new developments in instrumentation, bonding chemistries of capillaries, and applications on the analysis of proteins and their building-block components.

Although numerous examples of capillary electrophoresis separations of micro- and macromolecules can be cited (3-9), the most troublesome (and probably the application most commonly used), is the separation and analysis of proteins, peptides and amino acids. Table I shows a comprehensive view of the literature regarding the analysis of these substances, which are biologically the most diverse of all biological compounds, serving a vast array of functions.

The need for high-resolution protein separations has become more important due to the recent revolution in molecular biology. Typically, the recovery of an expressed protein from

TABLE I. Analysis of Proteins, Peptides, and Amino Acids by Capillary Electrophoresis

Analyte	Mode of CE	Detection System	Reference
Dansyl amino acids	Open-tubular	Fluorescence	10-14
Fluorescamine derivatized dipeptides	Open-tubular	Fluorescence	10
Leucine enkephalin vasotocin dipeptides	Open-tubular	Mass spectrometry	14
Lysozyme cytochrome c ribonuclease chymotrypsinogen horse myoglobin	Open-tubular	Fluorescence	15
Egg white lysozyme peptides	Open-tubular	Fluorescence	16
Chicken ovalbumin tryptic peptides	Open-tubular	Fluorescence	17
Phenylthiohydantoin amino acids	Open-tubular	UV	18
Human transferrin Human hemoglobin	Packed-tubular	UV	19
Cewl, hhcc, bprA, wsmm, esmm, hhm, dhm, dsmm,cewc, beca, bmlb, bmla, ceo	Open-tubular	UV	20
Horse myoglobin β-lactoglobulin A β-lactoglobulin B swm, hca, bca	Open-tubular	UV	21

Continued on next page

Table I. *Continued*

Analyte	Mode of CE	Detection System	Reference
D-L amino acids	Packed-tubular	UV	22
Human growth hormone	Packed-tubular	UV	22
α-Lactalbumin β-lactalbumin trypsinogen pepsin	Packed-tubular	UV	23
Rabbit hemoglobin	Packed-tubular	UV	4
OPA-amino acids	Open-tubular	Fluorescence	5
Myoglobin and myoglobin fragments	Open-tubular	UV	23
Synthetic peptides	Open-tubular	UV	24
Lysozyme trypsinogen myoglobin β-lactoglobulin A β-lactoglobulin B	Open-tubular	UV	25
Hirudin (thrombin-specific inhibitor)	Open-tubular	UV	5
Ggqa, ggea, ggda, wa, we, wg, ggra, wgg, wf	Open-tubular	UV	6
Untreated amino-acids, Dipeptides	Open-tubular	Electrochemistry	7,27
L-dihydroxy-phenylalanine	Open-tubular	Electrochemistry	26,27
Dipeptides	Open-tubular	UV	28

Continued on next page

Table I. *Continued*

Analyte	Mode of CE	Detection System	Reference
Phycoerythrin	Open-tubular	UV	29
Enolase β-amylase	Packed-tubular	UV	29
Chicken lysozyme β-lactoglobulin A β-lactoglobulin B rabbit parvalbumin hcc, hhm	Open-tubular	UV	28
Cytochrome c proteins	Open-tubular	UV	28
Neuropeptides	Open-tubular	UV	30-38
Prolyl 4-hydroxylase β-subunit peptides	Open-tubular	Fluorescence	39
Glycine, wsmm, carbonic anhydrase, β-lactoglobulin A, β-lactoglobulin B	Open-tubular	Fluorescence	40
Putrescine	Open-tubular	Fluorescence	41
Monoclonal antibodies	Open-tubular	Fluorescence	42,43

Abbreviations used here: cewl, chicken egg white lysozyme; hhcc, horse heart cytochrome c; bprA, bovine pancreas ribonuclease A; wsmm, whale skeletal muscle myoglobin; esmm, equine skeletal muscle myoglobin; hhm, horse heart myoglobin; dhm, dog heart myoglobin; dsmm, dog skeletal muscle myoglobin; cewc, chiken egg white conalbumin; beca, bovine erythrocytes carbonic anhydrase; bmlA, bovine milk β-lactoglobulin A; bmlB, bovine milk β-lactoglobulin B; ceo, chicken egg ovalbumin; swm, sperm whale myoglobin; hca, human carbonic anhydrase; bca, bovine carbonic anhydrase; hcc, horse cytochrome c.

tissue culture media or fermentation broths is difficult because host cell contaminants and artifacts of the recombinant product must be removed.   Artifacts arising from translation errors, improper folding, premature termination, incomplete or incorrect post-translational modification, and chemical or proteolytic degradation during purification all contribute to the production of polypeptide species with structures similar to the desired native polypeptide. Therefore, a high-resolution method such as capillary electrophoresis would be useful for monitoring biosynthetic fidelity and protein purity during the production of recombinant proteins. For example, it would be useful to separate peptides which differ only in one aminoacid or if the location of the same amino acid in the sequence is different. Similarly, it would be useful in the characterization of closely related proteins, such as isoenzymes and immunoglobulins.

The use of capillary electrophoresis as an analytical tool has been quite successful in the separation of a few small molecular weight proteins and many peptides obtained from commercial sources, most probably highly purified. However, *in cellulo*, proteins are usually associated with multimolecular complexes which are known to participate in essential cellular processes such as DNA replication, DNA recombination, and protein synthesis. Furthermore, other important biological processes that require protein complexes for activity include cellular motion, catalysis of metabolic reactions, regulation of biochemical processes, transport of micro- and macromolecules, and the structural maintenance of cells and the cellular matrix. In addition, the disruption of normal processes by viral infection produces virus-encoded multimolecular protein complexes, including the partially assembled precursors of the mature virus. Therefore, these complex protein-macromolecules may present a problem (for their separations) when using untreated fused-silica capillaries due to the adsorption of many proteins onto the walls of the capillary. Since the performance of any analytical technique is characterized in terms of accuracy, precision, reproducibility and dynamic range, many changes have to be made to the system in order to optimize the performance of capillary electrophoresis for the analysis of peptides and proteins.

For small peptides, separation efficiencies in excess of one million theoretical plates have been demonstrated (13,20,44).

Separation efficiencies for large proteins are more common in the hundred thousand theoretical plates. In comparison with gas chromatography, supercritical fluid chromatography, and liquid chromatography, capillary electrophoresis is the best separation technique from the point of view of molecular weight range of applicability. In the same column, it is possible to separate species ranging in size from free amino acids to large proteins associated with complex molecular matrices. In addition, from the detection standpoint, high-performance liquid chromatography is proven to provide better concentration sensitivity. On the other hand, capillary electrophoresis can provide better mass sensitivity.

As an instrumental approach to conventional electrophoresis, capillary electrophoresis offers the capability of on-line detection, micropreparative operation and automation (6,8,45-47). In addition, the *in tandem* connection of capillary electrophoresis to other spectroscopy techniques, such as mass spectrometry, provides high information content on many components of the simple or complex peptide under study. For example, it has been possible to separate and characterize various dynorphins by capillary electrophoresis-mass spectrometry (33). Therefore, the combination of CE-mass spectrometry (CE-MS) provides a valuable analytical tool useful for the fast identification and structural characterization of peptides. Recently, it has been demonstrated that the use of atmospheric pressure ionization using Ion Spray Liquid Chromatography/ Mass Spectrometry is well suited for CE/MS (48). This approach to CE/MS provides a very effective and straightforward method which allow the feasibility of obtaining CE/MS data for peptides from actual biological extracts, i.e., analysis of neuropeptides from equine cerebral spinal fluid (33).

Peptide mapping studies, generated by the cleavage of a protein into peptide fragments, must be highly reproducible and quantitative. Several electropherograms of protein digests have been obtained when chicken ovalbumin was cleaved by trypsin

(17), β−subunit of prolyl 4-hydroxylase cleaved by *Staphylococcus aureus* strain V8 protease (39), egg white lysozyme by trypsin (16), myoglobin and hemoglobin cleaved by trypsin (48), β-lactoglobulin A cleaved by *Staphylococcus* V8 protease (24), and recombinant interferon by trypsin (see Figure 1). Since capillary electrophoresis can provide high mass sensitivity, ultra-high efficiency and nanoliter sample injection, it also provides an excellent tool for the characterization of proteins when comparing peptide mapping, especially if the amount of material is difficult to obtain. For example, it could be quite useful for the identification of mutations in certain proteins which are characteristic of detrimental diseases (such as genetic diseases), or in the identification of site-specific protein modifications. In addition, it can be used as a quality control measure for recombinant protein products.

Routinely, common chemical and enzymatic techniques are used to obtain protein fragments. Unfortunately, when enzymatic digestion techniques and nanograms quantities of proteins are used, the method become ineffective due to dilution and reduced enzymatic activity. An alternative approach to overcome this problem is the use of proteolytic enzymes immobilized to a solid support and a small-bore reactor column. Using trypsin immobilized to agarose, tryptic digests of less than 100 ng of protein can be reproducible obtained (49).

The major concerns that are general to the use of all capillary electrophoresis systems for the separation of proteins and their building-block components are (a) choosing columns; (b) buffer solution compatibility with the system; and (c) the selection of the hardware.

CAPILLARY COLUMN

The heart of any chromatographic and electrophoretic system is the column. Preparation of capillary columns requires specific modifications, including bonding chemistries. Although one can

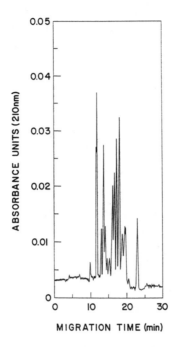

Figure 1. Electropherogram of Tryptic Digest Derived from Interferon. About 1 mg/ml of recombinant interferon (Hoffmann-La Roche, Inc.), was submitted to proteolytic digestion using 40 μl of a 1 mg/ml trypsin solution in 0.05 M Tris-acetate buffer, pH 7.5. The enzymatic digestion was carried out for 16 hr at 37°C. Approximately 5 nl of the protein digest was then separated by capillary electrophoresis using an untreated fused-silica column (75 μm x 100 cm), filled with 0.05 M sodium tetraborate buffer, pH 8.3. The peptides were monitored at 210 nm. Other experimental conditions are described in reference 8.

certainly prepare one's own capillary tube using various chemicals and cross-linked polymers, it is unlikely that most people will do so. Adoption of capillary electrophoresis in protein separations is dependent on the commercial availability of high quality crosslinking materials, packed columns, and surface-modified capillary columns that have been specifically designed for protein separations. At the present time commercially prepared capillary columns are only available as naked or surface untreated columns.

Although several intents were made to used Pyrex borosilicate glass columns, or teflon columns for use as the separation system in capillary electrophoresis (15,44), the most successful and commonly used capillary column today is the one made of vitreous silica. Except for the superior ultraviolet transparency of fused silica, Pyrex borosilicate glass and fused silica capillaries behave alike for use in capillary electrophoresis. Teflon, while also having good ultraviolet transparency, exhibits a poorer thermal conductivity than either Pyrex or fused silica, and thus has a greater tendency to overheat. The fused silica capillary column is available from various sources (for example, Scientific Glass Engineering, Austin, Texas, and Polymicro Technologies, Phoenix, Arizona). The fused-silica quartz capillary is optically compatible with ultraviolet as well as fluorescence detection.

BUFFER SOLUTION IN CAPILLARY ELECTROPHORESIS

In general terms, capillary electrophoresis is the electrophoretic separation of a substance from (usually) a complex mixture within a narrow tube filled with an electrolyte solution which is normally an aqueous buffer solution. Although one example of separation performed in a totally non-aqueous solution has been reported (50), neutral and slightly basic buffer solutions are generally used. Small tubes dissipate heat efficiently and prevent disruption of separations by thermally driven convection currents. Therefore, capillary electrophoresis can use relatively large electric fields to separate the components in very small samples rapidly and effectively.

In open-tubular capillary electrophoresis, a buffer-filled capillary is generally suspended between two reservoirs that contain the same buffer. A strong electroosmotic flow carries solutes, usually without regard to charge, from the positive end to the negative (that is, grounded electrode) end. In addition to the electroosmotic flow, electrophoresis also occurs. As a result, the components in the injected sample separate on bases of differences on their electrophoretic mobilities. In most cases, however, the rate of electrophoretic migration is slower than the rate of electroosmotic flow. Consequently, all species in the injected sample normally travel in one direction (the direction of the electroosmotic flow), allowing detection of positive, neutral, and negative species as they pass specific points along the capillary tubing.

Although capillary electrophoresis is a powerful technique for the separation of ionic compounds, some isomeric or closely related ionic compounds are not always successfully separated. Several attempts to improve selectivities are being carried out in various laboratories, as has been done for high-performance liquid chromatography. These attempts are usually through modification of the buffer components and conditions, and through modification of the capillary surface.

One of the important modifications of the system is the change in pH of the buffer. The pH of the buffer is the most critical factor for selectivity (18,51), because the protonation state of compounds having ionizable groups depends on the pH. The power of this approach is evident in the following example: the close related oxygen isotopic benzoic acids have been easily separated under an appropriate pH condition by means of an ionization control technique (52). The net charge of a substance and consequently the electrophoretic mobilities of zwitter ionic and multi-ionizable compounds such as amino acids, peptides and proteins, vary significantly with the pH (24,28). Other improvements in the separation of compounds by capillary electrophoresis, has been the addition of organic solvents to the buffer, such as methanol, propanol, and acetonitrile (27,44,53).

Additives that specifically interact with an analyte component are also very useful in altering the electrophoretic mobility of that component. For example, the addition of copper(II)-L-histidine (12) or copper(II)-aspartame (54) complexes to the buffer system allows racemic mixtures of derivatized amino acids to resolve into its component enantiomers. Similarly, cyclodextrins have proven to be useful additives for improving selectivity. Cyclodextrins are non-ionic cyclic polysaccharides of glucose with a shape like a hollow truncated torus. The cavity is relatively hydrophobic while the external faces are hydrophilic, with one edge of the torus containing chiral secondary hydroxyl groups (55). These substances form inclusion complexes with guest compounds that fit well into their cavity. The use of cyclodextrins has been successfully applied to the separation of isomeric compounds (56), and to the optical resolution of racemic amino acid derivatives (57).

Other modifications of capillary electrophoresis techniques, described below, have been adopted for the improvement of separation of several substances, including proteins, peptides and aminoacids. Furthermore, new procedures are also being developed to avoid the adsorption of proteins and peptides onto the walls of the capillary and consequently improving their selectivities.

One of the first steps in modifying the performance of capillary electrophoresis was the deactivation of silica groups of the capillary column by physically coating the capillary wall with methylcellulose (58,59), as well as via silane derivatization (10,44,60). Presently, many other changes have been carried out either to the capillary surface or addition of chemical agents to the separation buffer (see Table II), including manipulation of

TABLE II. Improvements on the Capillary Electrophoresis Separation of Proteins, Peptides and Amino Acids by Deactivation of the Silica Surface or by Addition of Chemical Agents to the Separation Buffer

Deactivating Agent	Chemical Agent	Other Conditions	Reference
Methylcellulose	-	-	29,58,59
-	Cyclodextrin	-	56,57
Glycophase	-	-	15
-	-	Low pH aqueous buffers	24,28
[(Methacryloyloxy) propyl]trimethoxysilane /1-vinyl-2-pyrrolidinone	-	-	28
-	-	High pH aqueous buffers	20
-	Morpholine	-	53
Trimethylchloro- silane	-	-	12,54,61
-	Detergents	-	54-56

the electric charge on the proteins and the silica capillary wall to prevent adsorption by Coulombic repulsion (20).

As mentioned above, the electroosmotic flow provides another means of transporting solutes, including neutral ones, through the capillary column. Differences in the viscous drag of neutral solutes, primarily, as a result of size differences, can provide for their separation. However, these differences are usually very small, and consequently, capillary electrophoresis is not very useful for separating structurally similar neutral compounds. In general terms, electrophoresis in free solutions is a separation method based solely on the difference in electrophoretic mobilities of charged species. It is not surprising, in principle, that electrically neutral compounds, of which the electrophoretic mobilities are essentially zero, cannot be separated. However, in order to make capillary electrophoresis effective for separating neutral compounds, addition of a surfactant to the electrophoretic buffer solution is necessary. The surfactant (at a concentration above its critical micelle concentration), effectively makes available a mechanism for separating neutral compounds. The use of these ionic micelles has been developed by Terabe and coworkers (63,64).

Charged micelles are subject to electrophoretic effect and therefore migrate with a different velocity from the surrounding aqueous phase. Micelles then act as a chromatographic phase, which may correspond to the stationary phase in conventional chromatography, by solubilizing the neutral molecules into their hydrophobic core. The separation is thus based on the differential distribution of the solute molecules between the electroosmotically-pumped aqueous phase and the micellar phase, which is moved by the cumulative effects of electroosmosis and electrophoresis. This technique is termed micellar electrokinetic chromatography (65) and also called micellar electrokinetic capillary chromatography (MECC) (66).

Micellar electrokinetic chromatography has been proven to be a highly efficient separation method for neutral analytes (63,67,68), neutral and charged compounds (69-72), and ionized compounds (67,73) including PTH-amino acids (18).

In certain cases, non-detergent substances act as micelles. For example, a cyclodextrin derivative containing ionizing groups has been employed as a substitute for a charged micelle (56). In this case, the cyclodextrin derivative acts as a chromatographic phase as it is in the case of the micelle, although it constitutes a homogeneous solution. The solute is partitioned between the hydrophobic region of the cyclodextrin and the aqueous phase. This technique (of using ionizable cyclodextrin derivatives) also can be classified as electrokinetic chromatography (65) and will separate neutral molecules very effectively.

CAPILLARY ELECTROPHORESIS INSTRUMENTATION

Several instruments have been developed in various laboratories since the late 1970s. Currently, several companies have introduced the capillary electrophoresis commercially (for example Microphoretic Systems, Sunnyvale, California; Bio-Rad, Richmond, California; Applied Biosystems, Inc., Foster City, California; and Beckman Instruments, Inc., Palo Alto, California). Although the instruments have many practical features for the separation and analysis of analytes, several new features need to be incorporated (for routine use) by protein chemists.

One of the most important features of the capillary electro-phoresis instrument is the detection system. Several of the detection methods commonly used in high-performance liquid chromatography have been somewhat adapted for capillary electrophoresis. Currently, the most popular detection methods developed for capillary electrophoresis are summarized in Table III.

Despite the availability of several detectors that have been successfully adapted for capillary electrophoresis, only ultraviolet and noncoherent light fluorescence are currently configured in the capillary electrophoresis instruments. Presently, the most common way to enhance detection of

TABLE III. Detection Methods Used in Capillary Electrophoresis

----

1. ABSORPTION
   a) Ultraviolet
   b) Visible

2. FLUORESCENCE
   a) Noncoherent light fluorescence
   b) Coherent light fluorescence or laser-induced fluorescence
   c) Epillumination fluorescence microscopy

3. ELECTROCHEMICAL
   a) Conductivity
   b) Potentiometric
   c) Amperometric

4. RADIOMETRIC

5. SPECTROSCOPY
   a) CE-Mass Spectrometry
   b) CE-Fourier Transform Infrared*
   c) CE-Raman*
   d) CE-Ultraviolet*

----

*Techniques under investigation.

analytes is by conjugating the analyte under study with a fluorescence tag. The fluorescent conjugate can then be detected using fluorescence techniques (see Table III, and 3,5,8,15,38,40, 41,44,54,74-79). Dovichi and co-workers (76) have been able to improve detection of amino acids derivatized with fluorescein isothiocyanate when detected with laser-induced fluorescence. The detection limits that can be reached with this technique are at the level of subattomole. Nevertheless, fluorescent detectors currently used still have limitations in maximixing sensitivity due to a large amount of light scattering which occurs before light reaches the detector. A solution to this problem has been overcome by the use of a fluorescence microscope (38,77-79). The capillary column was placed on the plate of a microscope equipped with a mercury lamp. The fluorescence was produced by epillumination through a chromatic beam splitter that reflected radiation of less than 420 nm and refracted radiation above 420 nm. An objective-condenser focused the UV beam on the capillary. The emitted visible light was focused through the lenses either on a film, a photodetector, or the retina of the observer. The major advantages of the fluorescence microscope then are: 1) visualization of the zones of the different analytes; 2) good estimation of the electroosmotic flow; 3) pictures can be taken or a video tape record made of the zones; 4) improvement of fluorescence detection sensitivity when compared with conventional fluorescence detectors adapted to capillary electrophoresis; and 5) accurate quantitation of fluorescence analytes by adapting a photomultiplier to the microscope.

In general, an ideal capillary electrophoresis instrument is composed of the following basic components: 1) an autosampler or autoloader, containing buffer and sample reservoirs; 2) a high-voltage power supply; 3) a fused-silica column; 4) an on-column and/or an off-column detector; 5) a recorder and/or integrator; 6) a microprocessor-controlled motor(s) and power supply system; and 7) an appropriate data handling system. A fraction collector connected to the grounded terminal of the capillary is employed when sample isolation and collection is considered (see Figure 2). The use of multiple capillaries for the loading and collection of larger quantities of proteins or peptides is also necessary, and consequently, since more current is consumed and greater heat is generated, a thermoregulated capillary compartment is strongly recommended. The equipment can be either modular (i.e., assembled of single components) (Figure 2) or a completely integrated apparatus (see Figure 3).

The autosampler, autoinjector, or autoloader can accomodate multiple vials capable of holding disposable microcentrifuge tubes or any other suitable sample container. The sample holder should be designed preferably in a conic shape and able to sustain small volumes, from normally 500 μl to 1.5 ml, to as little as 1 or 2 μl of sample. Because small sample volumes may evaporate if exposed for a long period of time to ambient temperature, it is recommended to have a container with a cap having a tiny hole for access to the electrode (high-voltage terminal) and the capillary column. Alternatively, the sample container may be capped at all times except when an injection occurs. An automatic system will remove the cap during the injection time and then place it back to its original position after the running time has ended. Both methods, manually and automated, will protect the sample from accidental contamination and from concentration due to total evaporation of the sample. Ideally, the autosampler should have interchangeable turntables (in addition to multiple vials) to accomodate sample containers of different sizes.

The introduction of the samples onto the capillary column can be carried out by either displacement techniques or electrokinetic migration. Three methods of displacement or hydrostatic injection are available: a) direct injection, or pressure; b) gravity flow, or siphoning; and c) suction. The electrokinetic injection method arose from findings that electroosmosis act like a pump (80). Both methods have advantages and disadvantages. For example, a bias has been reported in electrokinetically injected

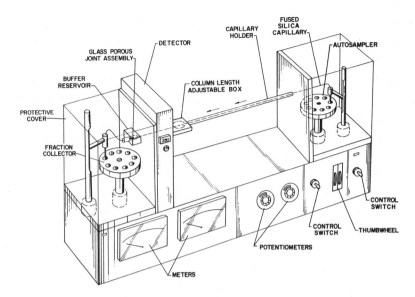

Figure 2. A Schematic Diagram of an Automatic Modular Capillary Electrophoresis Apparatus with On-column Detection. (Reproduced with permission from Ref. 71. Copyright 1988 Academic Press.)

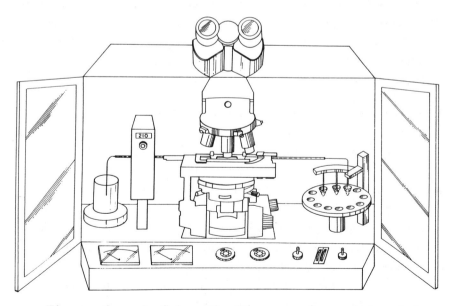

Figure 3. A Schematic Diagram of an Automated Integrated Capillary Electrophoresis Apparatus.

quantitative capillary electrophoresis analysis (81). Some researchers have assumed that this method delivers a representative sample into the capillary, but under certain conditions it does not. On the hand, hydrostatic injection can produce more quantitative results, although quantitative reproducibility still is a problem. The sample is introduced into the capillary column by using one of the injection methods described above, with the assistance of an autosampler. Currently, various models of autosamplers have been designed, including the use of a multiple sample holder-turntable or a 96-well microtiter plate as a sample holder (6,8,45,46).

Commercially available regulated high-voltage direct current (d.c.) power supplies are of three types: a bench top rack-mount unit with multiple features, a compact modular-type unit with unidirectional field-polarity capability, and a compact modular-type unit with bidirectional reverse field-polarity containing a built-in relay. These units can be operated either manually or through an actuated microprocessor-controlled system. High-voltage power supplies can be purchased from various sources (for example Hipotronics, Inc., Brewster, New York; Glassman High Voltage, Inc., Whitehouse Station, New Jersey; Bertan High Voltage, Hickville, New York; and Spellman High Voltage Electronics Corporation, Plainview, New York).

The need for a reverse-field polarity power supply is at least two-fold: 1) It permit a complete spectral analysis of the substance under study. By reversing field polarity, the substance zones can be run forward and backward in front of the detector as many times as needed. Incremental changes as small as 1- or 2-nm in wavelength can be used to maximize instrumental sensitivity, thus allowing coverage of the entire spectral range. In fact, this feature provides the same functions as a diode array detector, albeit somewhat slower. Proteins and peptides have almost identical spectral characteristics, however, when other functional groups are attached to it, is possible to observe more than one maximal absorbance peak. For example, in the case of horseradish peroxidase two maximal absorbance peaks are observed, one at approximately at 280 nm and another one at approximately 400 nm, due to the presence of a heme moiety bound to the protein molecule. The system described here allows more extensive investigation of a sample across a wider range of wavelengths at lower cost and higher sensitivity. 2) It facilitates the change between open- and packed-tube capillary electrophoresis. Under normal conditions in open-tube capillary electrophoresis, the direction of the electroosmotic flow of buffer moves from the high-voltage terminal to the grounded terminal. However, if changes in the composition and/or of the buffer is altered, it is then possible to control the intensity of the electroosmotic flow and the direction of migration of the analytes in the system. For example, replacement of the buffer by a polymeric matrix (such as agarose or acrylamide), or the addition of certain chemical substances, can reduce or suppress the electroosmotic flow completely (22,82). Therefore, the system can then be made operational by either the control of the electroosmotic flow or by the control of electrophoresis. Under the control of electro-phoresis, positive molecules will migrate toward the negative terminal, and negative molecules will migrate toward the positive terminal. Consequently, the field polarity must be changed according to the need. Furthermore (in capillary electrophoresis) the length of the capillary column needs to be changed. The presence of polymeric matrices within the capillary column will generate greater heat (difficult to be dissipated as in normal buffer-containing columns) that may cause deleterious problems, such as the melting of agarose or the formation of bubbles. A solution to this problem is to use capillary columns of shorter lengths, i.e., 10- to 25-cm long. In general, shorter columns are routinely used in packed-tube capillary electrophoresis. Also, a thermoregulated compartment to maintain the system under a desired temperature and a system for column length are strongly recommended (see below).

Another important feature of a power supply is the availability of a programming voltage time in the unit. Improvement in the resolution of proteins (when separated by a voltage gradient), has been reported (28).

A device can be implemented to the instrument in order to rejuvenate or recycle the capillary column after multiple injections of samples. The recycling or cleanup procedure of the column would ensure good performance, as well as prolonging the life of the capillary column.

The most common cause of early failure of the capillary column is lack of care in solvent preparation and inadequate cleanup of biological samples. Buffers and other additives may contain dissolved impurities (which may be retained in the capillary column), resulting in a slowing-down of the flow of the buffer in the column to virtually no detectable motion of the fluid. The first step in avoiding this problem is the use of deionized and triply distilled water in the preparation of the buffers, and the use of solvents and solutes of high purity (solvents of HPLC grade or spectroscopy grade, and crystallized solutes). Filtering the final mobile phase (prior to use) through an appropriate micropore (0.22 or 0.45 μm) filter, and degassing the buffers and solvents will remove particulated matter from the solutions and will minimize bubble presence during the separation of the analytes. The presence of bubbles will sever the normal passage of current resulting in a complete stoppage of the electrical circuit and consequently the buffer flow. Solutions can be degassed by several methods (for example, by vacuum, ultrasonic bath, or inert gas such as helium), but a degassing system that can be attached to the instrument is particularly advantageous. We have found that during the time that one particular sample is separated, two other adjacent samples can be degassed. This process can be carried out in microcentrifuge tubes using very thin microtubing that carries a controlled amount of inert gas. After this "on-line" degassing process (see

Figure 4), the autosampler moves the next microcentrifuge tube into position for sample loading onto the capillary. At this stage, the new sample (and buffer) has already been degassed, allowing a new cycle of sample to be examined, and a new set of samples to be degassed. No contamination of new sample occurs when this method of alternating buffer solution (after every sample to be analyzed) is used.

Column failure may also arise from the gradual accumulation of particulate material that adheres to the walls of the capillary column, usually originating from "sticky" proteins and other macromolecules such as lipids and other substances. The source of strongly retained substances are commonly tissue or cell homogenates, and biological fluids. When using serum, for example, the performance of the capillary column will diminish approximately after the fifth or sixth injection and the migration time of the analytes under study will change significantly after multiple injections of serum, or other biological fluids. Similarly, deterioration in performance can also be observed when other biological fluids are examined. One solution to the problem (although impractical) is to replace the capillary column after a certain number of injections since commercially available fused-silica capillaries are inexpensive. However, because future improvements in the technology of capillary electrophoresis includes the use of bonding phases on the surface of the capillary column, the price of the commercially coated-columns will (probably) increase significantly. It will be difficult to then make the capillary column a disposable item. A solution to this problem can be overcome with the use of a stainless-steel or teflon tee device connected by ferrules especially made for the capillary column. This tee device is developed as a cleanup system when attached to a vacuum pump and a fluid trap. The key in rejuvenating a contaminated column is by using a cleaning procedure that can be carried out by purging with various solutions added in a sequential fashion: phosphoric acid, deionized water, potassium hydroxide, deionized water, and finally aspirating and priming the capillary with buffer (from vials in the autosampler). All fluids exit the teflon port into the fluid trap. The capillary column is then ready for a new separation cycle. The cleaning cycle may be carried out manually each time changes in the performance of the separation occurs, or it may be programmed automatically with a microprocessor-controlled system (every cycle) if necessary (83). After the cleaning procedure (and before the injection of a new sample), it is important to run the system with plain buffer for a short period of time, or until a base line is reached.

A general recommendation is that samples injected onto capillary columns should be as free as possible of contaminating material (to minimize interferences in detection and to prevent unnecessary adsorption of sample components on the column). Optimizing sample volume and sample concentration for optimal electropherographic resolution, and to avoid band broadening and tailing resulting in reduced separation efficiencies of the analytes under study, is additionally recommended. Typically, a capillary column of 75 μm x 100 cm (total capacity or separation volume of the column is 4.4 μl) can be loaded with 1 to 20 nanoliters of volume sample, and a concentration of proteins and peptides in the range of approximately 10 μg to 5 mg/ml of solution, depending upon the method of detection to be used. Keeping careful operating conditions will prolong the life of a capillary column. The life of a capillary column should be remarkable good for at least 100 injections or more. Due to the small capacity volume of the narrow-bore capillary column, maintaining the two ends of the column submerged in liquid at all times is recommended. If the ends of the column are exposed to open air for a large period of time, evaporation will occurs and salts and aggregates will be deposited within the capillary column making the column permanently ineffective.

Another important component of the capillary electrophoresis instrumentation is the cassette-cartridge design (Figure 5). This modular component has three functions: 1) protection of the capillary column, 2) the accomodation of capillary columns of

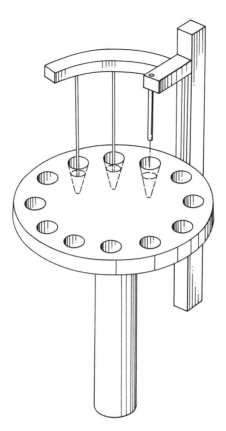

Figure 4. A Schematic Diagram of an On-line Degassing System.

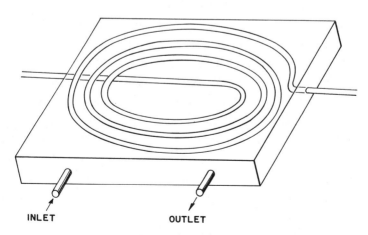

INLET   OUTLET

Figure 5. A Schematic Diagram of a Cassette-Cartridge Device.

various sizes and compositions, and 3) regulation of the cooling system. Although the capillary column seems to be very durable (because of the externally coated polyimide polymer), a region of the column (approximately 1- to 2-cm) is burned-off allowing it to become transparent. This transparent region is centered and aligned to allow the passage of a light beam for detection and quantitation of the analytes under study. Careful handling conditions of the capillary is recommended at this stage (since the new uncoated region become very fragile). The accomodation of various sizes and chemistries of the capillaries makes the system very convenient for the preparation of different capillary columns in advance, as well as making the system very simple (for column replacement). The cartridge has channels, allowing the passage of a thermoregulated fluid.

Finally, the dream of every protein chemist is not only to separate components as homogeneously as possible, but also to collect significant quantities in order to further perform analyses. A fraction collector, and the technology to isolate measurable amounts of proteins is essential to the capillary electrophoresis instrument. In principle, the process of fraction collection in capillary electrophoresis is fundamentally different from that in liquid chromatography. The end of the capillary must stay in contact with the buffer solution (and the terminal electrode) during the fraction collection in order to maintain a closed electrical circuit. If the capillary column is transferred from one collecting reservoir to another, the electric field is discontinued (during the transference), the electroosmotic flow of the system is interrupted, and consequently the migration of solutes is stopped. The process is regenerated when the capillary (and the terminal electrode) is again placed back in contact with a new buffer-containing reservoir. This sequence of events is ideal to transfer the capillary from fraction to fraction and to discretely collect separated zones of analytes. A similar system of fraction collection has been used by Jorgenson and co-workers (84). One limitation to this approach is the large dilution that a protein or peptide undergoes. The collecting reservoir can usually hold a volume ranging from 5- to 25 µl-buffer. Another limitation is the need to collect fractions in a solid-support surface rather than in a liquid-containing vessel. The collection in a membrane-type surface is especially useful in certain applications of the diagnostic industry or in special research techniques. In our laboratory (8) we have designed a fraction collector capable of collecting fractions (after sample components have been separated on the capillary column) that does not require the end of the capillary to stay in contact with the buffer solution (and the terminal electrode). This fraction collector is based on the use of a porous glass assembly, designed for off-column electrochemical detection (7,73). The grounding of the electric system (in this design) is carried out at the end of the separation capillary and at the beginning of the detection capillary. Under these conditions, no collecting reservoir is needed at the terminal of the capillary in order to maintain a normal current flow. Therefore, the substances to be further analyzed (or collected) continue to be pushed along the capillary. As a consequence, microdrops of fluid are formed at the end of the tip of the capillary column which can then be collected in a buffer-filled microcentrifuge tube, a dry vessel, or directly into a membrane-type solid support. This segment of the entire capillary is not subjected to the deleterious effects of high voltages (the effects of the high-voltage electric field are eliminated). Therefore, it is possible to assemble a detector in close proximity to the tip of the capillary (i.e., fiber optic sensor), quite useful for monitoring the separated analytes at the precise time of collection.

Using a single capillary to collect a separated component may present a problem to the user (from the point of view of quantity). Currently, capillary electrophoresis is used primarily for analytical tests. However, two approaches have been performed to use capillary electrophoresis as a micro- or semi-preparative technique. One approach is done by increasing sample load and detector response by arranging capillaries in bundles (85). The ideal instrument should be configured to sustain several capillaries. By using four to eight capillaries (75-100 μm i.d., 100 cm length) in open-tubular operation, we have collected purified proteins and peptides in nanograms quantities (85), which could be a sufficient peptide sample for microsequencing analysis. In addition, with capillary bundles, a device can be constructed that is capable of holding multiple detectors (one for every capillary column) and consequently capable of performing several simultaneous electrophoretic separations. A second micro-preparative operational approach has been the use of wider diameter columns (150-200 μm i.d., 10-25 cm length, packed with polymeric matrices (i.e., acrylamide) yielding a recovery of approximately 1 μg of analyte (9,22).

Analytical or micro-preparative operation in capillary electrophoresis does not appear to alter the integrity of the holomeric structure of proteins, and consequently all their biological activities maintained. This is particularly true if appropriate cooling conditions are used. For example, using capillary electrophoresis separation, three enzymes (bovine pancreatic α-chymotrypsin (84), chick embryo and human placental prolyl 4-hydroxylase (86), have been separated and collected maintaining more than 95% of their enzymatic activities. Similarly, specific antibodies produced against enzymatically active purified human placental prolyl 4-hydroxylase have also been separated and collected (using capillary electrophoresis) maintaining more than 97% of their immunological activities (87).

The advancement of modern biochemistry and developments in micro- and macromolecular separations have been intimately linked. Capillary electrophoresis offers major advantages over other separation techniques, including speed and resolving power. The potential of capillary electrophoresis seems so vast that it will significantly complement the technology of high-performance liquid chromatography. However, because of unique characteristics of capillary electrophoresis, it will also replace many existing technologies. Currently, protein chemists as well other scientists, are rapidly discovering the many uses of this powerful technique. Despite all the progress, capillary electrophoresis is still in the early stages of becoming a routine technology among scientists, and faces many improvements in the years to come.

LITERATURE CITED

1. Electrophoresis: A Survey of Techniques and Applications; Part A: Techniques (Deyl, Z., ed.); Elsevier, Amsterdam, The Netherlands, 1979.

2. Gaal, O.; Medgyesi, G.A; Verczkey, K. Electrophoresis in the Separation of Biological Macromolecules. Wiley-Interscience, Chichester, U.K., 1980.

3. Jorgenson, J.W.; Rose, D.J.; Kennedy, R.T. Am. Lab. 1988, 20, 32.

4. Compton, S.W.; Brownlee, R.G. BioTechniques 1988, 6, 432.

5. Gordon, M.J.; Huang, X.; Pentoney, Jr., S.L.; Zare, R.N. Science 1988, 242, 224.

6. Brownlee, R.G.; Compton, S.W. Am. Biotechnol. Lab. 1988, 6, 10.

7. Ewing, A.G.; Wallingford, R.A.; Olefinowicz, T.M. Anal. Chem. 1989, 61, 292A.

8. Guzman, N.A.; Hernandez, L.; Hoebel, B.G. Biopharm Manufact. 1989, 2, 22.

9. Karger, B.L.; Cohen, A.S.; Guttman, A. J. Chromatogr. 1989, 492, 585.

10. Jorgenson, J.W.; Lukacs, K.D. Anal. Chem. 1981, 53, 1298.

11.   Green, J.S.; Jorgenson, J.W.   J. High Resol. Chromatogr.
      Chromatogr. Comun. 1984, 7, 529.

12.   Gassmann, E.; Kuo, J.E.; Zare, R.N.   Science 1985, 230, 813.

13.   Walbroehl, Y.; Jorgenson, J.W.   J. Microcolumn Separat.
      1989, 1, 41.

14.   Smith, R.D.; Olivares, J.A.; Nguyen, N.T., Udseth, H.R.   Anal.
      Chem. 1988, 60, 436.

15.   Jorgenson, J.W.   New Directions in Electrophoretic Methods,
      Jorgenson, J.W.; Phillips, M., eds., (American Chemical
      Society, Washington, D.C.), pp. 182-198, 1987.

16.   Jorgenson, J.W.; Lukacs, K.D.   J. High Resol. Chromatogr.
      Chromatogr. Commun. 1981, 4, 230.

17.   Jorgenson, J.W.; Lukacs, K.D.   J. Chromatogr. 1981, 218, 209.

18.   Otsuka, K.; Terabe, S.; Ando, T.   J. Chromatogr. 1985, 332,
      219.

19.   Hjertén, S.   J. Chromat. 1985, 347, 191.

20.   Lauer, H.H.; McManigill, D.   Anal. Chem. 1986, 58, 166.

21.   Jorgenson, J.W.   Anal. Chem. 1986, 58, 743A.

22.   Cohen, A.S.; Paulus, A.; Karger, B.L.   Chromatographia 1987,
      24, 15.

23.   Cohen, A.S.; Karger, B.L.   J. Chromatogr. 1987, 397, 409.

24.   Grossman, P.D.; Wilson, K.J.; Petrie, G.; Lauer, H.H.   Anal.
      Biochem. 1988, 173, 265.

25.   Jorgenson, J.W.   Trends Anal. Chem. 1984, 3, 51.

26.   Wallingford, R.A.; Ewing, A.G.   J. Chromatogr. 1988, 441,
      299.

27. Wallingford, R.A.; Ewing, A.G. Anal. Chem. 1989, 61, 98.

28. McCormick, R.M. Anal. Chem. 1988, 60, 2322.

29. Hjertén, S.; Zhu, M.-D. J. Chromatogr. 1985, 327, 157.

30. Guzman, N.A.; Hernandez, L. Sixth International Symposium on Isotachophoresis and Capillary Zone Electrophoresis, Vienna, Austria, 21-23 September 1988.

31. Hernandez, L.; Hoebel, B.G.; Guzman, N.A. The 196th National Meeting of American Chemical Society, Los Angeles, California, 25-30 September 1988. Abstract ANYL-182.

32. Guzman, N.A.; Hernandez,L.; Hoebel, B.G. Annual Meeting of the Society for Neurosciences. Toronto, Canada, 13-18 November 1988.

33. Henion, J.; Muck, W.; Huang, E. 40th Pittsburgh Conference and Exposition on Analytical Chemistry and Applied Spectroscopy. Atlanta, Georgia, 6-10 March 1989. Abstract 1195.

34. Guzman, N.A.; Hernandez, L.; Advis, J.P. Current Research in Protein Chemistry (J.J. Villafranca, Ed.), Academic Press, New York, 1990, in press.

35. Advis, J.P.; Hernandez, L.; Guzman, N.A. Peptide Research, 1989, in press.

36. Guzman, N.A.; Advis, J.P.; Hernandez, L. Third Symposium of The Protein Society. Seattle, Washington, July 29-August 2 1989. Abstract T-197.

37. Advis, J.P.; Hernandez, L.; Guzman, N.A. The 1989 International Chemical Congress of Pacific Basic Societies. Honolulu, Hawaii, 17-22 December 1989.

38. Advis, J.P.; Guzman, N.A.; Fourteenth International Symposium on Column Liquid Chromatography. Boston, Massachusetts, 20-25 May 1990.

39. Guzman, N.A.; Gonzalez, C.L.; Trebilcock, M.A.; Sixth International Symposium on HPLC of Proteins, Peptides, and Polynucleotides. Baden-Baden, West Germany, 20-22 October 1986. Abstract 1009.

40. Green, J.S.; Jorgenson, J.W. J. Chromatogr. 1986, **352**, 337.

41. Rose, D.J., Jr.; Jorgenson, J.W. J. Chromatogr. 1988, **447**, 117.

42. Guzman, N.A.; Hernandez, L. Techniques in Protein Chemistry (T. Hugli, Ed.), Chapter 44, pp. 456, Academic Press, New York, 1988.

43. Guzman, N.A.; Berck, C.M.; Advis, J.P. First International Conference on Human Antibodies and Hybridomas. Lake Buena Vista, Florida, 18-20 April, 1990.

44. Jorgenson, J.W.; Lukacs, K.D. Science 1983, **222**, 266.

45. Deml, M.; Foret, F.; Bocek, P. J. Chromatogr. 1985, **320**, 159.

46. Honda, S.; Iwase, S.; Fujiwara, S. J. Chromatogr. 1987, **404**, 313.

47. Kennedy, R.T.; Jorgenson, J.W. Anal. Chem. 1988, **60**, 1521.

48. Huang, E.; Henion, J. First International Symposium on High Performance Capillary Electrophoresis. Boston, Massachusetts, 10-12 April 1989. Abstract T-P-124.

49. Cobb, K.A.; Liu, J.; Novotny, M. 40th Pittsburgh Conference and Exposition on Analytical Chemistry and Applied

Spectroscopy. Atlanta, Georgia, 6-10 March 1989. Abstract 1422.

50. Walbroehl, Y.; Jorgenson, J.W. J. Chromatogr. 1984, **315**, 135.

51. Fujiwara, S.; Honda, S. Anal. Chem. 1986, **58**,1811.

52. Terabe, S.; Yashima, N.; Tanaka, N.; Araki, M. Anal. Chem. 1988, **60**, 1673.

53. Fujiwara, S.; Honda, S. Anal. Chem. 1987, **59**, 487.

54. Gozel, P.; Gassman, E.; Michelsen, H.; Zare, R.N. Anal. Chem. 1987, **59**,44.

55. Szejtli, J. Cyclodextrin and their Inclusion Complexes, Academic Press, Budapest, 1982.

56. Terabe, S.; Ozaki, H.; Otsuka, K.; Ando, T. J. Chromatgr. 1985, **332**, 211.

57. Guttman, A.; Paulus, A.; Cohen, A.S.; Grinberg, N.; Karger, B.L. J. Chromatogr. 1988, **448**, 41.

58. Hjertén, S. Chromatogr. Rev. 1967, **9**,122.

59. Herren, B.J.; Shafer, S.G.; Alstine, J.V.; Harris, J.M.; Snyder, R.S. J. Colloid Interface Sci. 1987, **115**, 46.

60. Chang, S.H.; Gooding, K.M.; Regnier, F.E. J. Chromatogr. 1976, **120**, 321.

61. Mosher, R.A. First International Symposium on High Performance Capillary Electrophoresis. Boston, Massachusetts, 10-12 April 1989. Abstract T-L-7.

62. Nielsen, R.G.; Sittampalam, G.S.; Rickard, E.C. First International Symposium on High Performance Capillary Electrophoresis. Boston, Massachusetts, 10-12 April 1989. Abstract M-P-123.

63. Terabe, S.; Otsuka, K.; Ichikawa, K.; Tsuchiya, A.; Ando, T. Anal. Chem. 1984, 56, 111.

64. Terabe, S,; Otsuka, K.; Ando, T. Anal. Chem. 1985, 57, 834.

65. Terabe, S. Trends Anal. Chem. 1989, 8, 129.

66. Burton, D.E.; Sepaniak, M.J.; Maskarinec, M.P. Chromatographia 1986, 21, 583.

67. Cohen, A.S.; Terabe, S.; Smith, J.A.; Karger, B.L. Anal. Chem. 1987, 59, 1021.

68. Otsuka, K.; Terabe, S.; Ando, T. Nippon Kagaku Kaishi 1986, 950.

69. Otsuka, K.; Terabe, S.; Ando, T. J. Chromatogr. 1985, 348, 39.

70. Burton, D.E.; Sepaniak, M.J.; Maskarinec, M.P. J. Chromatogr. Sci. 1986, 24, 347.

71. Fujiwara, S.; Iwase, S.; Honda, S. J. Chromatogr. 1988, 447, 133.

72. Nishi, H.; Tsumagari, N.; Kakimoto, T.; Terabe, S. J. Chromatogr. 1989, 465, 331.

73. Wallingford, R.A.; Ewing, A.G. Anal. Chem. 1988, 60, 258.

74. Tsuda, T.; Kobayashi, Y.; Hori, A.; Matsumoto, T.; Suzuki, O. J. Chromatogr. 1988, 456, 375.

75. Wright, B.W.; Ross, G.A.; Smith, R.D. J. Microcolumn Separat. 1989, 1, 85.

76. Cheng, Y.-F.; Dovichi, N.J. Science 1988, 242, 562.

77. Hernandez, L.; Escalona, J.; Marquina, R.; Guzman, N.A. First International Symposium on High Performance Capillary

Electrophoresis.   Boston,   Massachusetts,   10-12   April,   1989.
Abstract   T-P-115.

78. Hernandez, L.; Marquina, R.; Escalona, J.; Guzman, N.A.   J.
Chromatogr.   1989, in press.

79. Hernandez, L.; Joshi, N.; Escalona, J.; Guzman, N.A.   Second
International Symposium on High Performance Capillary
Electrophoresis.   San Francisco, California, 29-31 January
1990.

80. Petrorius, V.; Hopkins, B.J.; Schieke, J.D.   J.   Chromatogr.
1974, **99**, 23.

81. Huang, X.; Gordon, M.J.; Zare, R.N.   Anal.   Chem.   1988, **60**,
375.

82. Altria, K.D.; Simpson, C.F.   Chromatographia   1987, 24, 527.

83. Guzman, N.A.; Hernandez, L.; Advis, J.P.   The 41st
Pittsburgh Conference & Exposition on Analytical
Chemistry and Applied Spectroscopy.   New York, New York,
5-8 March 1990.   Abstract 706.

84. Rose, D.J.; Jorgenson, J.W.   J.   Chromatogr.   1988, 438, 23.

85. Guzman, N.A.; Regnier, F.E.   First Symposium of The Protein
Society.   San Diego, California, 9-13 August 1987.   Abstract
823.

RECEIVED December 20, 1989

Chapter 2

Applications of Capillary Zone Electrophoresis to Quality Control

R. G. Nielsen and E. C. Rickard

Lilly Research Laboratories, Eli Lilly and Company, Indianapolis, IN 46285

Performance parameters were established for the use of
capillary zone electrophoresis (CZE) in the characterization of biosynthetic human growth hormone (hGH).
CZE distinguished between the trypsin digest fragments
of hGH and corresponding fragments produced from most
of its chemically similar impurities and degradation
products. Linearity, precision, and sensitivity were
established for intact hGH and its major desamido degradation product; they were comparable to values
observed with other separation methods. In addition,
the effect of eluting solvent pH on peak shape was
determined. Finally, the electrophoretic flow velocity
was determined as a function of pH. It was found that
CZE has important applications with respect to verifi-
cation of identity (trypsin digests) and to determination of purity (detection and quantitation of
impurities) of hGH. It can be considered an important
adjunct to conventional chromatographic procedures.

Pharmaceutical products are characterized for identity, potency,
and purity. In addition, their stability is followed with time
to observe the formation of degradation products and the concomitant loss in potency. A diverse variety of analytical techniques are used to evaluate these parameters in the bulk drug
(active ingredient) and drug product (formulated product). For
example, spectroscopic examinations of the intact material may be
used to confirm identity. Potency (amount of active ingredient or
biological activity in a dosage unit) and purity (lack of impurities
and degradation products) are frequently evaluated by separative
techniques such as reversed-phase high performance liquid chromatography (RP-HPLC). However, the structural complexity of proteins
requires that more than one method be used to quantitate the amount
of material and to verify its identity and integrity. We evaluated

0097-6156/90/0434-0036$06.00/0
© 1990 American Chemical Society

the role of capillary electrophoresis in a free solution or open
tubular mode (capillary zone electrophoresis, CZE) for determination of the identity and purity of biosynthetic human growth
hormone (hGH).

The hGH molecule consists of a single peptide strand with two
intrachain disulfide bridges. The molecular weight and isoelectric
point of hGH are 22,250 and 5.2, respectively (1). It has been
extensively characterized by conventional techniques that include
RP-HPLC, size exclusion chromatography, anion exchange chromatography, sodium dodecyl sulfate polyacrylamide gel electrophoresis,
isoelectric focusing electrophoresis, and amino acid analysis
(2-5). Impurities in hGH result from incomplete removal of related
substances during manufacture and purification. Degradation
products form by deamidation at Asn-149 and/or Asn-152, by oxidation of the thioether at methionine-14, and by formation of a
non-covalent dimer (5-6). Techniques used to establish the
identity of hGH must distinguish these chemically similar
molecules; methods that monitor the potency and purity must dis-
criminate between the desired bioactive species and the related
substances.

Confirmation of the identity for each lot of hGH is performed
with a limited set of techniques to verify that it is identical to a
reference standard lot. One of the most powerful procedures involves comparison of the chromatographic fingerprint obtained from
a proteolytic enzyme (e.g., trypsin) digest of the sample to a
similar fingerprint obtained from the reference standard. When
hGH is digested with trypsin under conditions that preserve the
disulfide linkages, it is cleaved into 19 peptides that reflect the
primary amino acid sequence and the configuration of the two disulfide bonds (6-7). Since it is easier to detect changes in these
small peptides than in the intact molecule, an identical fingerprint
of the digests from two samples is relatively conclusive proof of
the correct structure (i.e., identity) of the intact molecule.

Determinations of potency and purity of the bulk drug substance
and of the formulated drug product are other major concerns of
pharmaceutical companies. Determination of purity requires quantitation of impurities and degradation products in the presence of
much larger quantities of the main component. For example, the
chemical purity of hGH, like many other recombinant-DNA derived
products, is greater than 95%, and individual contaminants are
present at levels of 0.5-2% or less. (The biological purity is
typically about 98%; i.e., only about 2% of the material is biologically inactive.) Thus, the determination of purity is inherently
related to the specificity and sensitivity of the methods used to
detect contaminants, whereas methods used to measure potency require
high specificity but relatively low sensitivity. Techniques such
as RP-HPLC, that have high separative power, are easily automated,
and produce quantitative data, are frequently used to quantitate
impurities present at the 0.1-5% level. Other techniques that also
possess high separative power, such as slab gel electrophoresis,
are much less useful because they cannot be easily quantitated and
they are very labor intensive.

CZE has been widely used for separations due to its high separative power, simplicity of operation, ease of quantitation, and ability to perform rapid, automated analyses (8-11). It has been applied to separation and characterization of diverse species, especially polypeptides and proteins (12-16). We recently reported separations of biosynthetic human insulin (BHI) and human growth hormone (hGH) from closely related species that could originate as impurities or as degradation products (15). That paper demonstrated that the results for the quantitation of the desamido-A21 BHI degradation product in human insulin by CZE were equivalent to results obtained from reversed-phase high performance liquid chromatography (RP-HPLC) and polyacrylamide gel electrophoresis. Later papers reported quantitation of BHI related compounds, the use of CZE for peptide mapping of a hGH digest (16), identification of peaks in the hGH digest (17-18), and optimization of the CZE separation conditions for the hGH digest (18). Another recent paper reports extensive data on linearity, precision, and limit of detection for several polypeptides and proteins (19). This paper will describe how CZE can be used to monitor identity and purity for quality control of hGH.

Experimental

Reagents and Materials. Biosynthetic hGH, [desamido-Asn-149] hGH, [didesamido-Asn-149, Asn-152] hGH, and [Met-sulfoxide-14] hGH were obtained from Eli Lilly and Co. (Lilly Research Laboratories, Indianapolis, IN). Morpholine, 2-[N-morpholino]ethanesulfonic acid (MES), and 3-[N-morpholino]propanesulfonic acid (MOPS) were purchased from Fisher Scientific Co. (Pittsburgh, PA); tricine was purchased from Sigma Chemical Co. (St. Louis, MO). Tris (hydroxymethyl) aminomethane (TRIS) was purchased from Bio-Rad Laboratories (Richmond, CA). Trypsin (TPCK, 267 units/mg protein, 98 percent protein) was purchased from Cooper Biomedical, Inc. (Malvern, PA). TRIS-acetate buffer was prepared by adjusting the pH of a 0.05 M TRIS solution to pH 7.5 with acetic acid. Reagent grade water obtained from a Milli-Q purification system from Millipore Corp. (Bedford, MA) was used to prepare all solutions. All other reagents were analytical grade and were used without further purification. Polyimide-coated, fused silica capillaries, 50 micrometer internal diameter and 360 micrometer outside diameter, were purchased from Polymicro Technologies, Inc. (Phoenix, AZ).

Methods. The trypsin digestion was carried out according to reported methods using non-reducing conditions so that both the correct amino acid sequence and the presence of the correct disulfide linkages could be confirmed (6). Aliquots of the digest mixture were frozen (-20°C) for use at a later time. The thawed digest mixture was injected directly. The concentration of the analyte in all studies was about 2 mg/mL total protein or 90 uM for each digest fragment except for fragments 17, 18, and 19 that will be present at lower concentrations since fragments 17 and 19

derive from cleavage of fragment 18. Thus, the nominal loading is
about 20 ng of intact species or about 1 picomole of each fragment
in a digest mixture.

The mobile phase used for the CZE separations of the trypsin
digests was 0.1 M tricine and 0.02 M morpholine adjusted to pH
8.15. The mobile phase composition for the linearity and precision
data was 0.01 M tricine, 0.0058 M morpholine, and 0.02 M NaCl adjusted to pH 8.0. The mobile phase buffers used to determine peak
shape and electrophoretic mobility vs. pH all contained 0.02 M
NaCl and 0.01 M of the buffer (pH 3.0, citrate; pH 4.0, formic
acid; pH 5.0, acetate; pH 6.0, MES; pH 7.0, MOPS; pH 8.0, tricine;
pH 9.0, borate; and pH 10.0, glycine) and were adjusted to the
indicated pH with acid or base. The column was rinsed with mobile
phase between injections or successively with 0.1 M sodium hydro-
oxide and mobile phase when the mobile phase composition was
changed.

With the exception of the digests, CZE experiments were performed using the same instrumentation as previously described (15)
except that both CZE instruments now include vacuum injection
devices and a constant temperature environment. Electropherograms
of the hGH digests were obtained with a model 270A Applied
Biosystems Inc. (Santa Clara, CA) instrument. Sample volumes were
estimated from the Poiseuille equation. Sample was introduced by
applying vacuum to a capillary that was approximately 100 cm in
length (total) and 80 cm between the point of injection and the
detector. Separation conditions were: 30 kV applied voltage and
25° or 30°C. The components were detected by UV absorbance at
200 nm. Analog data were collected directly from the absorbance
detector on an in-house centralized chromatography computer system
based on a Hewlett-Packard model 1000 minicomputer that has
storage, manipulation, and graphics capabilities.

Results and Discussion

The selectivity achieved in CZE separations is determined by
differences in electrophoretic mobilities of the analytes.
Mobility in the open tubular mode is predominantly related to
charge, shape, and size of the analyte as well as to the properties
of the eluting solvent (9). Thus, the separation of impurities and
degradation products that have similar shapes and sizes will be
determined principally by differences in their net charge (20).
Adjustment of pH to a value near the midpoint of the isoelectric
point (pI) range of the analytes will tend to maximize their net
charge differences, provided that the proteins are soluble near
their pI. These criteria were met at pH 8 for the intact species.
However, a more complex optimization strategy was followed for the
heterogeneous mixture of peptide fragments produced by trypsin
digestion (18).

Identity. Peaks in the electropherogram of a tryptic digest of
hGH were identified by a combination of spiking and amino acid
analysis (17-18). After extensive optimization, CZE was shown to
be as good or better in separative power when compared to RP-HPLC
and faster than the corresponding RP-HPLC separation. However,
like RP-HPLC, CZE still could not separate all of the digest fragments with any one separation condition (18). Fortunately, it is
rarely necessary to separate all of the fragments since the likely
impurities and degradation products will affect only specific
fragments.

        Figure 1 shows the optimized separation of the tryptic digest
for authentic hGH. Note that under certain conditions, fragments
11 and 14 can partially degrade to form new species labelled 11* and
14*, respectively. Figure 2 shows the comparision between digests
of hGH and [desamido-Asn-149] hGH. Residue Asn-149 is located in
fragment 15 of the tryptic digest. It contains 13 residues from
position 146 to 158 that have the sequence NH_2-Phe-Asp-Thr-Asn-Ser-
His-Asn-Asp-Asp-Ala-Leu-Leu-Lys-COOH. It is apparent that fragment
15 of the desamido derivative has shifted to a longer migration
time, consistent with the formation of an additional negative charge
from the conversion of the asparagine amide to an aspartic acid
carboxylate. The change of -1 in the net charge at pH 8 reduces
the mobility of fragment 15 since the negative electrode is located
at the detector end of the capillary. This trend is even more
dramatic in Figure 3, which shows the electropherograms of hGH
and [didesamido-Asn-149, Asn-152] hGH. Fragment 15 of the dides-
amido derivative which contains residues 149 and 152 (see above)
has a change of -2 in the net charge at pH 8 compared to the
corresponding fragment in hGH. Finally, the electropherograms of
hGH and [Met-sulfoxide-14] hGH are given in Figure 4. Residue
Met-14 is located in fragment 2 of the tryptic digest which
consists of the sequence NH_2-Leu-Phe-Asp-Asn-Ala-Met-Leu-Arg-COOH
(positions 9 to 16). In this case, we cannot see a shift in
fragment 2, the location of the oxidized methionine, since there is
no difference in net charge and very little difference in size,
shape, or other parameters that would affect their electrophoretic
migration rates. RP-HPLC is the only technique that has been
successful in separation of the sulfoxide; its presumed separation
mechanism involves a change in hydrophobicity (6).

Potency and purity. Chromatographic methods are frequently used to
measure the quantity of a single active species which is equated to
the potency for most drugs. Related substances may be quantitated
by similar procedures, usually incorporating gradient elution (for
ion exchange or RP-HPLC methods) to insure that all species are
observed. The related substance assay defines the purity of the
material. The performance characteristics of these methods are
crucial for the correct evaluation of potency and purity.

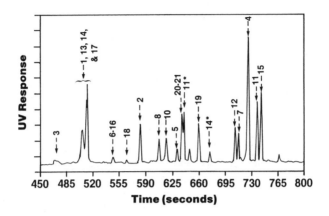

Figure 1. Electropherogram of hGH digest; elution buffer, 0.1 M tricine, and 0.02 M morpholine, pH 8.15; injection volume about 9 nL.

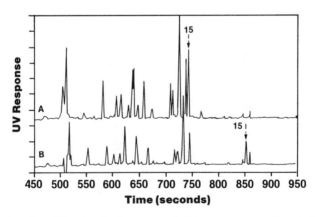

Figure 2. Electropherograms of digests, experimental conditions given in Figure 1. Location of fragment 15 as indicated. A. hGH, and B. [Desamido-Asn-149] hGH.

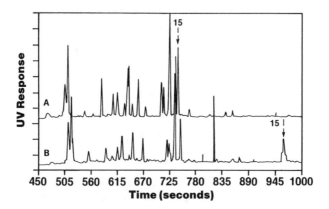

Figure 3. Electropherograms of digests, experimental
conditions given in Figure 1. Location of fragment 15
as indicated. A. hGH, and B. [Didesamido-Asn-149,
Asn-152] hGH.

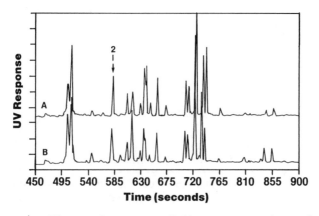

Figure 4. Electropherograms of digests, experimental
conditions given in Figure 1. Location of fragment 2 as
indicated. A. hGH, and B. [Met-sulfoxide-14] hGH.

Method validation includes determination of performance characteristics such as selectivity (which determines accuracy), linearity, precision, and sensitivity (limit of detection). This work evaluated linearity, precision, and sensitivity for specific CZE separation conditions; selectivity was reported previously (15). Factors that contribute to assay imprecision by affecting peak shape (such as the pH of the mobile phase) or migration velocity (pH effects on the electrophoretic velocity) were evaluated also.

Linearity. The linearity of response was evaluated by varying the injection volume (changing the sample introduction time while keeping a constant 1.0 mg/mL sample concentration). The injection volume covered the range of 2.44 nL to 22.0 nL, corresponding to a 1- to 9-second sample introduction time using a vacuum equivalent to 3.75 inches of mercury. The peak areas obtained from a mixture of hGH and [desamido-Asn-149] hGH were plotted vs. injection volume. The linearity was excellent, with observed correlation coefficients of 0.9998, 0.9997, and 0.9998 for the hGH, desamido, and total peak areas, respectively. However, a positive y-intercept was noted for all plots. Subsequent experiments demonstrated that the positive y-intercept was due to an offset in the timing circuit. That is, the sample introduction time was about 0.3 seconds longer than the nominal value, an amount that approximately corresponded to the offset of 1.3 nL in the x-intercept. No attempt has been made to extend the range of linearity for hGH. However, previous work (16) demonstrated a wide linear range for biosynthetic human insulin that covered a factor of 64 in sample concentration and a factor of 9 in injection time. These results are supported by other work (19) that reported good linearity for area vs. injection time (volume) and area vs. concentration over wide ranges.

Precision. The precision of multiple injections was evaluated using triplicate injections (at each nominal volume) of the same hGH/desamido hGH mixture that was used for the linearity study (Table I). The percent of desamido hGH in the mixture was calculated as the ratio of its peak area to the total peak area; raw peak areas must be corrected for differences in migration velocity as previously described (15).

There are several conclusions that can be drawn from these results. First, the peak areas have a pooled RSD of about 5%, a value which is much better than that expected for densitometric scanning of electrophoresis gels. Second, note that the standard deviation of the peak area measurements was essentially independent of the volume injected. Thus, the relative standard deviation of the peak area dramatically decreases as the injection volume increases. An injection volume of at least 10 nL is required to obtain good precision (2-3%, excluding a single poor replicate

for the desamido area at 10 nL); larger volumes appear to be even
better, although large volumes adversely affect resolution (10).
The observed precision is comparable to the values we previously
reported for biosynthetic human insulin (16). It also is similar to
independent results obtained using a totally automated system (2.9%
RSD) and much better than that reported for manual injection (11.8%
RSD), both using a hydrodynamic injection technique (21). Finally,
the observed precision for the percent desamido, which is really an
area ratio similar to what would be obtained by comparison to an
internal standard, is excellent for the 10-nL or larger injections.
Although the data are insufficient to make a definitive conclusion,
it suggests that the observed error is comparable to that obtained
from many chromatographic techniques. It also suggests that one of
the predominant sources of error is imprecision in the injection
volume. The error in injection volume was recently characterized
(19). They also reported approximately 1-3% RSD in peak areas for
vacuum injection of various compounds.
    An alternative way of examining the above data is presented
in Figure 5, which shows the 3 individual replicates (points), the
average value (line), and the standard deviation (error bar) for
the percent desamido vs. sample volume. It can be observed that
the average percent desamido is invariant with injection volume,
as would be expected. The improvement in precision at high sample
volumes is obvious.
    The other aspect of precision is day-to-day reproducibility.
At this time, we have no data on this aspect of the assay performance. All of the data discussed above were obtained in sets run
on one day.

Sensitivity. Sensitivity or limit of detection refers to the
minimum amount or minimum concentration of analyte that may be
distinguished from a blank with the desired confidence level. It
is determined by the ratio of the maximum quantity of material that
can be injected onto the column to the minimum amount of material

Table I. Precision of Peak Areas and Desamido Content

| Volume | Peak Areas | | | | | | Percent Desamido Content | |
| | hGH | | DA-hGH | | Total | | | |
(nL)	SD	RSD	SD	RSD	SD	RSD	Ave	RSD
2.44	560	4.7%	1040	9.1%	1460	6.2%	48.6%	3.9%
4.88	1350	7.0%	520	2.8%	980	2.5%	49.5%	4.6%
9.77	710	2.0%	2870	8.1%	3470	4.9%	50.0%	3.2%
15.87	1620	3.0%	1490	2.8%	2760	2.6%	50.0%	1.3%
21.98	1680	2.3%	820	1.1%	2440	1.7%	49.9%	0.7%
Pooled	1270	4.2%	1580	5.8%	2400	4.0%	49.6%	3.1%

that can be detected reliably. The maximum amount of sample is limited by three factors: (1) sample solubility, (2) electric field inhomogeneities caused by changes in conductivity associated with differences between the sample or sample matrix and the separation buffer, and (3) loss of detector linearity for concentrated samples. For hGH and the conditions used in this work, field in- homogeneity (inferred from loss of resolution) occurs at about the 25- to 50-ng level; the other factors are less limiting. At the other extreme, the minimum amount of sample is determined by the sensitivity limit of the detector. For our conditions, this is probably less than about 0.1 ng. Using 0.1 and 25 ng as conserv- ative estimates of the minimum and maximum quantities, respect- ively, the dynamic range is 250-fold which corresponds to being able to observe 0.4% of an impurity in a 25-ng protein sample. Thus, UV detection at 200 nm provides adequate sensitivity for detection and measurement of hGH related substances when the sample solutions are about 1 mg/mL.

The quantity and volume of samples required for impurity de- termination by CZE are very small; probably less than 5 uL of volume is required for a well-designed injector, and only a few nanoliters (i.e., a few nanograms) are actually injected. However, it is experimentally simpler if that sample is present in a relatively concentrated solution, 0.05-2 mg/mL, when UV detection is being used. Our focus was not to achieve ultra-low detection limits such as might be required for trace level contaminants or for quantitation of trace levels of natural products. For those appli- cations, the most common approach has been the use of a laser-based detector, preferably combined with a fluorescent label on the analyte. With this combination, extremely low limits of detection can be achieved (9, 22-25).

Peak shape and electrophoretic mobility. Changes in the peak shape of hGH with elution conditions reflect interaction with other species in solution or with the wall. The change in the electro- phoretic mobility of hGH with pH is predominately related to changes in its charge. Either of these factors can affect the precision of peak area measurements (as well as the selectivity). Thus, the peak shape and electrophoretic mobility of hGH were examined over the range of pH 3 to pH 10 at unit intervals except for pH 5 where it was insoluble (the pH is close to the isoelectric point). Representative electropherograms are given in Figure 6. At low pH values, hGH is positively charged and migrates faster than the neutral marker, mesityl oxide. However, it migrates slower than the neutral marker at high pH values where it is nega- tively charged. Our preliminary data indicated acceptable peak shape in the neutral to moderately basic regions (pH 6 to 9) and relatively poorly shaped peaks in the moderately acidic regions. Since peak shape and selectivity were good near pH 8, the precision and linearity data were obtained at that pH.

The plot of electrophoretic mobility vs. pH is given in Figure 7. This plot is analogous to a charge-pH titration curve.

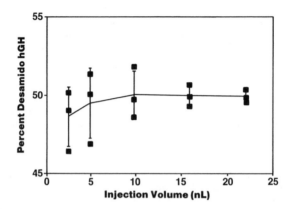

Figure 5. Percent [Desamido-Asn-149] hGH vs. injection
volume; elution buffer, 0.01 M tricine, 0.0058 M
morpholine, and 0.02 M NaCl, pH 8.0. Replicate values
(points), average value (line), and standard deviation
(error bars) as indicated.

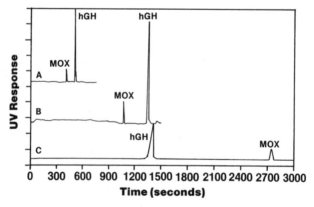

Figure 6. Electropherograms of hGH at selected pH values.
A. pH 9.0, B. pH 6.0, and C. pH 4.0. Buffer compositions are given in text; MOX is mesityl oxide.

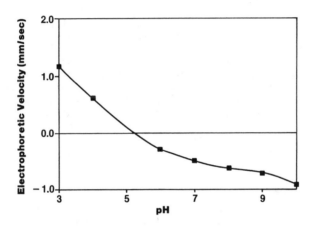

Figure 7. Electrophoretic velocity of hGH vs. pH. Buffer
compositions are given in text; velocity measured vs.
mesityl oxide.

A smooth curve is obtained which indicates a reasonable migration
velocity whose value is relatively independent of pH near pH 8.
Although the pH 5 value is missing as noted above, the interpolated
curve crosses the velocity axis at about pH 5.25, essentially
identical to the accepted pI value of 5.2 where the electrophoretic
velocity is expected to be zero. (However, the net velocity at all
pH values is towards the detector due to the presence of electroosmotic flow.)

Conclusion

Capillary electrophoresis has been demonstrated to be useful in
monitoring the identity and purity of hGH. CZE is capable of
discrimination between hGH and several closely related impurities
and degradation products, either as the intact species (previous
work) or as the trypsin digests (this work). Thus, it is an
important adjunct to conventional methods, such as RP-HPLC of
digests or conventional electrophoresis of intact proteins, for
the identification of hGH. CZE possesses adequate sensitivity to
monitor minor impurities; it is comparable to RP-HPLC with respect
to linearity and precision. Samples with a volume of at least 10
nanoliters will provide acceptable precision; those that contain
an internal standard will provide the best precision. The peak
area response is linear; it is possible to extend linearity by
increasing concentration as long as the contribution of the sample
and its matrix to field inhomogeneities is minimized. The work
shows that CZE has potential to be useful in the quality control
of proteins such as hGH.

Acknowledgments

We thank Drs. J. W. Jorgenson, R. M. Riggin, G. W. Becker, and
G. S. Sittampalam for helpful discussions and encouragement. We
acknowledge ISCO, Inc. for the donation of a detector. We are
grateful for the technical assistance given by D. S. LeFeber and
P. A. Farb.

Literature Cited

1. Atkins, L. M.; Miner, D. J.; Sittampalam, G. S.; Wentling,
   C. D. J. Assoc. Off. Anal. Chem. 1987, 70, 610-617.
2. Riggin, R. M.; Dorulla, G. K.; Miner, D. J. Anal. Biochem.
   1987, 167, 199-209.
3. Riggin, R. M.; Shaar, C. J.; Dorulla, G. K.; LeFeber, D. S.;
   Miner, D. J. J. Chromatogr. 1988, 435, 307-318.
4. Sittampalam, G. S.; Ellis, R. M.; Miner, D. J.; Rickard, E. C.;
   Clodfelter, D. K. J. Assoc. Off. Anal. Chem. 1988, 71, 833-838.
5. Becker, G. W.; Bowsher, R. R.; MacKellar, W. C.; Poor, M. L.;
   Tackitt, P. M.; Riggin, R. M. Biotech. Appl. Biochem. 1987,
   9, 478-487.

6. Becker, G. W.; Tackitt, P. M.; Bromer, W. W.; LeFeber, D. S.;
    Riggin, R. M. Biotech. and Applied Biochem. 1988, 10, 326-337.
7. Conova-Davis, E.; Chloupek, R. C.; Baldonado, I. P.; Battersby,
    J. E.; Spellman, M. W.; Basa, L. J.; O'Connor, B.; Pearlman, R.;
    Quan, C.; Chakel, J. A.; Stults, J. T.; Hancock, W. S.
    Amer. Biotechnology Laboratory May, 1988, 8-17.
8. Jorgenson, J. W. Anal. Chem. 1986, 58, 743A-758A.
9. Gorden, M. J.; Huang, X.; Pentoney, S. L.; Zare, R. N. Science
    1988, 242, 224-228.
10. Compton, S. W.; Brownlee, R. G. Biotechniques 1988, 6, 432-440.
11. Ewing, A. G.; Wallingford, R. A.; Olefirowicz, T. M.
    Anal. Chem. 1989, 61, 292A-303A.
12. Jorgenson, J. W.; Lukacs, K. D. Science 1983, 222, 266-272.
13. Lauer, H. H.; McManigill, D. Anal. Chem. 1986, 58, 166-170.
14. McCormick, R. M. Anal. Chem. 1988, 60, 2322-2328.
15. Nielsen, R. G.; Sittampalam, G. S.; Rickard, E. C.
    Anal. Biochem. 1989, 177, 20-26.
16. Grossman, P. D.; Colburn, J. C.; Lauer, H. H.; Nielsen, R. G.;
    Riggin, R. M.; Sittampalam, G. S.; Rickard, E. C. Anal. Chem.
    1989, 61, 1186-1194.
17. Nielsen, R. G.; Riggin, R. M.; Rickard, E. C. J. Chromatogr.,
    1989, 480, 393-401.
18. Nielsen, R. G.; Rickard, E. C. J. Chromatogr., submitted.
19. Moring, S. E.; Colburn, J. C.; Grossman, P. D.; Lauer, H. H.
    LC-GC 1990, 8, 34-46.
20. Grossman, P. D.; Wilson, K. J.; Petrie, G.; Lauer, H. H.
    Anal. Biochem. 1988, 173, 265-270.
21. Rose, D. J.; Jorgenson, J. W. Anal. Chem. 1988, 60, 642-648.
22. Rose, D. J.; Jorgenson, J. W. J. Chromatogr. 1988, 447,
    117-131.
23. Cheng, Y.; Dovichi, N. J. Science 1988, 242, 562-564.
24. Pentoney, S. L.; Huang, X.; Burgi, D. S.; Zare, R. N.
    Anal. Chem. 1988, 60, 2625-2629.
25. Christensen, P. L.; Yeung, E. S. Anal. Chem. 1989, 61,
    1344-1347.

RECEIVED January 17, 1990

Chapter 3

Analysis of Cyclic Nucleotides by Capillary Electrophoresis Using Ultraviolet Detection

Luis Hernandez[1], Bartley G. Hoebel[2], and Noberto A. Guzman[3,4]

[1]Department of Physiology, School of Medicine, Los Andes University, Merida, Venezuela
[2]Department of Psychology, Princeton University, Princeton, NJ 08540
[3]Protein Research Unit, Princeton Biochemicals, Inc., Princeton, NJ 08540

Capillary electrophoresis, a powerful high-efficiency high-resolution analytical technique, was used for the separation and characterization of cyclic-AMP, cyclic-GMP, and cyclic-IMP. Reproducibility, linear-ity, and spectral analysis were tested. The results shows that capillary electrophoresis is a reliable technique used to resolve and quantitate sub-picomole amounts of a mixture of cyclic nucleotides.

Cyclic nucleotides are purinic base derivatives with powerful biological activity. It is widely accepted that cyclic nucleotides mediate many of the intracellular biochemical events triggered by neurotransmitters and hormones (1,2). Therefore, the analysis of these compounds carries special relevance in biological sciences. A wide variety of techniques has been developed for cyclic nucleotide assays including binding to phosphokinase (3,4) or to antibodies (5); activation of enzymes (6); or separation techniques such as thin layer chroma-tography (7) and high-performance liquid chromatography (8-14). However, each of these techniques have some limitations, including the complexity of the assay or the volumes needed to reach a reasonable sensitivity. The emergence of capillary electrophoresis (CE) has gradually begun to solve problems in which the handling of low nanoliter samples and low concen-

[4]Current address: Roche Diagnostic Systems, Inc., 340 Kingsland Street, Nutley, NJ 07110–1199

trations (subfemtomole quantities) is necessary. In this technique, a high voltage electric field provides the driving force to move the chemicals (and their separation is performed as they move) through a small bore capillary tube (15-17). Capillary electrophoresis offers a high number of theoretical plates which improves resolution, and since it works with small volumes it might detect small (mass) amounts of cyclic nucleotides. Therefore, we explored the application of CE to the analysis of some cyclic nucleotides. Our long term goal is to combine brain perfusion techniques (such as push-pull and microdialysis) with CE (18-20) (to elucidate the chemical changes underlying brain functions). In the present paper we show that the resolution and analysis of cyclic nucleotides by CE with ultraviolet (UV) detection is feasible in the picomole range.

EXPERIMENTAL SECTION

Instrumentation. The CE apparatus used here was similar to the one previously described (17,21,22). It has a capillary bridging two reservoirs connected to a high-voltage power supply (Spellman High Voltage Electronics Corporation, Plainview, New York). It also has a computerized system for sample injection and analysis, using a modified on-column ultraviolet detection system (EM Science-Hitachi, Gibbstown, New Jersey). One reservoir is a 1 ml disposable microcentrifuge tube, and the other a 50 ml plexiglass beaker. The capillary column (externally coated with the polymer polyimide) used has the following dimensions: 75 µm i.d., 300 µm e.d., and 100 cm long (Scientific Glass Engineering, Austin, Texas). A small area of the capillary (about 1 cm) was stripped of the coating (by burning) at approximately 55 cm from the high voltage or injection terminal. This uncoated section was placed in an aligned position with the path of the ultraviolet light beam of the detector. The high voltage power supply provided constant voltage and variable current, applying 10 kV for 15 sec (for electrokinetic loading of the sample), and 22 kV for 30 min (for moving the sample through the capillary). The electrodes are platinum-iridium wires. The high voltage (positive) electrode and one end of the capillary are connected to a motorized (and computer-controlled) arm which lowered them into the vessel containing the sample. The field polarity between the ends of the capillary could be switched to run the zones backward and forward (repeatedly) for spectral analysis. The on-column detector was modified to accept the capillary rather than a cuvette. The modified system allowed rapid change of the capillary column while preserving its correct position within the optic chamber. Keeping careful operating conditions will prolong the life of a capillary column and will insure highly reproducible values of the samples to be analyzed. A cleaning procedure was used (after the analyses of ten samples), consisting of purging the capillary column with potassium hydroxide, rinsing with deionized water, and aspirating and priming the capillary with buffer (before a new cycle starts).

Reagents. Cyclic nucleotides (3',5'-cyclic adenosine mono-phosphate (c-AMP), 3',5'-cyclic guanosine monophosphate (c-GMP), and 3',5'-cyclic inosine monophosphate (c-IMP)); sodium tetraborate; hydrochloric acid; and potassium hydroxide were purchased from Sigma Chemical Company, St. Louis, Missouri). Millex disposable filter units (0.22 μm) were obtained from Millipore Corporation (Bedford, Massachusetts). Triply distilled and deionized water was used for the preparation of buffer solutions. Both buffers and samples were routinely degassed with helium after filtration (using microfilter units).

Methods. A 0.05 M sodium tetraborate buffer, pH 8.3, adjusted with 1 N HCl, was used as the electrophoretic buffer solution. Negative pressure from a vacuum pump was used to prime the capillary. This pump was temporarily connected to the capillary by means of a modified hypodermic needle and a piece of polyethylene tubing. After the filling of the capillary column, its two ends were immersed in the reservoirs, and nucleotide samples were analyzed by open-tubular free-zone capillary electrophoresis. Each nucleotide was tested for reproducibility, linearity, and stability (at room temperature). In addition, spectral analysis was performed. Reproducibility of the system was tested by injecting ten samples of the same solution. The peak height and the retention time were measured, averaged, and the dispersion of the values was calculated. An estimate percent error for each of these measurements was obtained by dividing their confidence limits (standard deviation) by each mean and the results multiplied by 100 (coefficient of variation). The confidence limits for peak height and migration time were calculated assuming a "t" distribution. Linearity was tested by injecting 6 or 7 different concentrations of each nucleotide, and a regression analysis was used (to estimate that the values fit to a straight line).   Stability at room temperature was tested by running each nucleotide for the first three hours, then every two hours over a six hour period   and at the 22nd hour.   The spectral analysis of ultraviolet (light) response was performed by alternating the field polarity to run the sample zone (containing the nucleotide) backward and forward through the detector at different wavelengths.   Finally, the three nucleotides were mixed together at equimolar concentrations and separated by capillary electrophoresis to test the resolution of this powerful technique.

RESULTS AND DISCUSSION

As shown in Table 1, the average and the dispersion of the absorbance, and the retention time of each nucleotide is compared.   The three graphs in Figure 1 show that the optical absorbances of the samples were linearly related to the concentration of the different nucleotides.   The regression analysis of the samples showed an almost perfect fit for linearity in the range of 40 μM to 4 mM concentration.   The fact that the slope of cAMP regression line was higher than the slope of c-GMP or c-IMP is probably due to their differences in light absorbance at the low UV range (for example, at 210 nm c-AMP absorbes more UV light than the other nucleotides).   The stability test showed that c-AMP and c-GMP were stable at room temperature (data not shown), but c-IMP degraded at a constant rate (Figure 2).   The spectral analysis shows that between 180- and 290-nm the three nucleotides have a bimodal absorption curve (Figure 3).   The wavelength range for maximal absorption was slightly different for the three nucleotides.   Cyclic-AMP absorbance reached a maximum between 200- and 210-nm, and another between 250- and 260-nm.   Cyclic-GMP absorbance was maximal between 180- and 190-nm, and between 240- and 250-nm.   Cyclic-IMP absorbance was maximal between 190- and 200-nm, and between 230- and 240-nm.   As a consequence the spectrum of c-AMP was displaced toward the right with respect to the spectra of c-GMP and c-IMP.   The mixture of the three nucleotides was resolved as shown in Figure 4.   The nucleotides migrated in the following order: c-AMP, c-GMP, and c-IMP.

The present results show that cyclic nucleotides can be analyzed by capillary electrophoresis and ultraviolet detection.   The three nucleotides (two of which have been detected in living organisms) showed different migration times and UV spectra.

TABLE 1. Reproducibility of Migration Time and Ultraviolet
Absorbance of Cyclic Nucleotides

Cyclic Nucleotide	Absorbance (AU x 10^2)	Percent Error	Migration Time (sec)	Percent Error
c-AMP	13.3 ± 0.3	2.3	1119 ± 3	0.3
c-GMP	8.1 ± 0.2	2.5	1163 ± 3	0.3
c-IMP	9.2 ± 0.3	3.3	1280 ± 3	0.2

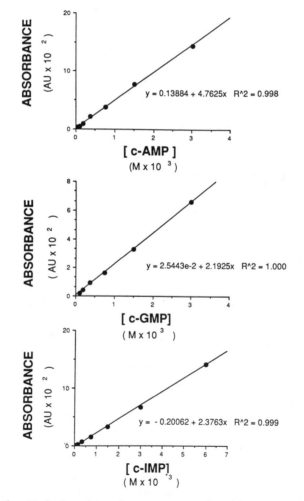

Figure 1. Relationship between nucleotide concentration and UV absorption (top curve: c-AMP, middle curve: c-GMP, and bottom curve: c-IMP).

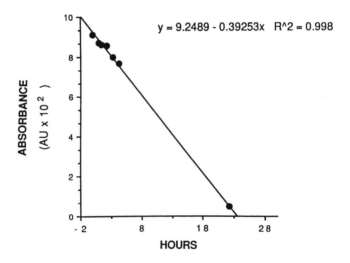

Figure 2. Time Course of Spontaneous Degradation of c-IMP.

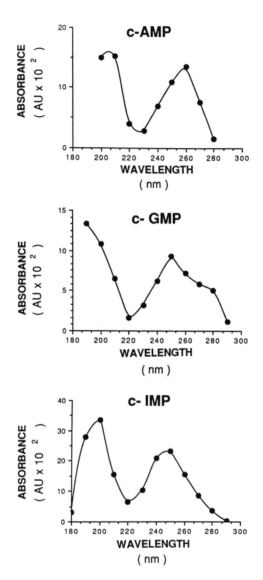

Figure 3. Ultraviolet Spectra of c-AMP (top curve), c-GMP (middle curve), and c-IMP (bottom curve).

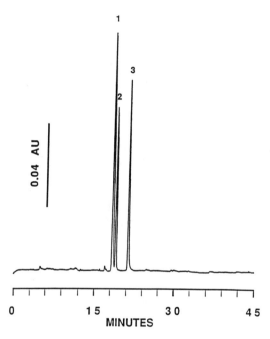

Figure 4. Electropherogram of a mixture of c-AMP (1), c-GMP (2), and c-IMP (3). Samples were monitored at 210 nm.

These features allow good separation and identification of these nucleotides. Capillary electrophoresis showed excellent linearity between 0 and 40 mM concentration. Considering that in the present conditions the capillary electrophoresis apparatus can load approximately 4 nl, we estimate that the assay is linear between 0 and 160 picomoles. The fact that the zone containing a particular nucleotide is not deformed after successive runs permits rapid spectral analysis in a mixture of nucleotides by reversing the field polarity. This is particular important for unstable nucleotides such as c-IMP.

LITERATURE CITED

1. Robinson, G.A.; Butcher, R.W.; Sutherland, E.W. Cyclic AMP (Sutherland, E.W., and Robinson G.A., Eds.), Academic Press, New York, 1971.

2. Swillens, S.; Dumont, J.E. Cell Regulation by Intracellular Signal (Swillens, S., and Dumont, J.E., Eds.), Plenum Press, 1982.

3. Brown, B.L.; Albano, J.D.M.; Ekins, R.P.; Sgherzi, A.M.; Tampion, W. Biochem. J. 1971, **121**, 561.

4. Gilman, A.G. Adv. Cyclic Nucleot. Res. 1972, **2**, 9.

5. Steiner, A.L.; Wehmann, R.E.; Parker, C.W.; Kipnis, D.M. Adv. Cyclic Nucleot. Res. 1972, **2**, 51.

6. Kuo, J.F.; Greengard, P. Adv. Cyclic Nucleot. Res. 1972, **2**, 41.

7. Edhem, I.; Das, I.; Debeller, J.; Hirsch, S.R. Biochem. Soc. Trans. 1986, **14**, 1151.

8. Brooker, G. Fed Proceed. 1971, **30**, 140.

9. Anderson, F.S.; Murphy, R.C. J. Chromatogr. 1976, **121**, 251.

10. Martinez-Valdez, H.; Kothari, R.M.; Hershey, H.V.; Taylor, M.V. J. Chromatogr. 1982, **247**, 307.

11. Schulz, D.W.; Mailman, R.B. J. Neurochem. 1984, **42**, 764.

12. Lin, L.; Saller, C.F.; Salama, A. J. Chromatogr. 1985, **341**, 43.

13. Braumann, T.; Jastorff, B.; Richter-Landsberg, C. J. Neurochem. 1986, **47**, 912.

14. Alajoutsijarvi, A.; Nissinen, E. Anal. Biochem. 1987, **165**, 128.

15. Jorgenson, J.W.; Rose, D.J.; Kennedy, R.T. Amer. Laborat. 1988, **20**, 32.

16. Gordon, M.J.; Huang, X.; Pentoney, S.L., Jr.; Zare, R.N. Science 1988, **242**, 224.

17. Guzman, N.A.; Hernandez, L.; Hoebel, B.G. BioPharm Manufact. 1989, **2**, 22.

18. Hernandez, L.; Stanley, B.G.; Hoebel, B.G. Life Sci. 1986, **39**, 2629.

19. Guzman, N.A.; Advis, J.P.; Hernandez, L. Third Symposium of the Protein Society. Seattle, Wahington, July 29-August 2, 1989.

20. Advis, J.P.; Hernandez, L.; Guzman, N.A. The 1989 International Chemical Congress of Pacific Basin Societies. Honolulu, Hawaii, December 17-22, 1989.

21. Guzman, N.A.; Hernandez, L. Techniques in Protein Chemistry (Hugli, T.E., Ed.), Chapter 44, pp. 456, Academic Press, 1989.

22. Guzman, N.A.; Hernandez, L.; Terabe, S. Separations in Analytical Biotechnology. ACS Symposium series (J. Nickelly, and C. Horvath, Eds.), American Chemical Society, Washington, D.C., this book.

RECEIVED December 20, 1989

Chapter 4

On-Column Radioisotope Detection for Capillary Electrophoresis

Stephen L. Pentoney, Jr.[1,3], Richard N. Zare[1], and Jeff F. Quint[2]

[1]Department of Chemistry, Stanford University, Stanford, CA 94305
[2]Advanced Development Unit, Beckman Instruments, Inc.,
Fullerton, CA 92634

Three on-line radioactivity detection schemes for
capillary electrophoresis are described. The first
detector system utilizes a commercially available
semiconductor device positioned external to the
separation channel and responding directly to impinging
γ or high-energy β radiation. The second detector
system utilizes a commercially available plastic
scintillator material and a cooled photomultiplier tube
operated in the photon counting mode. The third
detector system utilizes a plastic scintillator material
and two room-temperature photomultiplier tubes operated
in the coincidence counting mode. The system
performance and detector efficiency are evaluated for
each of the detection schemes using synthetic mixtures
of ^{32}P-labeled sample molecules. The detection limits
are determined to be in the low nanocurie range for
separations performed under standard conditions (an
injected sample quantity of 1 nanocurie corresponds to
110×10^{-18} moles of ^{32}P). The lower limit of detection
is extended to the sub-nanocurie level by using flow
(voltage) programming to increase the residence time of
labeled sample components in the detection volume. The
separation of ^{32}P-labeled oligonucleotide mixtures using
polyacrylamide gel-filled capillaries and on-line
radioisotope detection is also presented. When desired,
the residence time can be made almost arbitrarily long
by freezing the contents of the capillary, permitting
autoradiograms to be recorded. This last technique is
applied to gel-filled capillaries and provides a
detection sensitivity of a few DPM per separated
component, corresponding to subattomole amounts of
radiolabel.

[3]Current address: Advanced Development Unit, Beckman Instruments, Inc.,
Fullerton, CA 92634

0097–6156/90/0434–0060$08.50/0

The highly efficient separations afforded by capillary
electrophoresis (CE) are a direct result of employing extremely
narrow separation channels. Effective dissipation of heat generated
by the passage of electrical current through the separation medium
occurs only when the ratio of capillary inner surface area to
internal volume is sufficiently large (typically 10^4 to 10^5 m^{-1}).
For this reason capillary tubes with internal diameters ranging from
10 μm (1) to 200 μm have been selected for CE separations.
   Early in the development of capillary electrophoresis, it was
noted that the successful detection of separated sample components
present within the narrow confines of these capillary tubes posed a
major challenge (2). In response to this challenge, much research
has been directed toward the development of sensitive and selective
detectors for capillary electrophoresis. CE detector technology has
been largely borrowed from the field of high-performance liquid
chromatography (HPLC), especially from micro-column HPLC.
Successful extension of the various HPLC detection schemes to
capillary electrophoresis has generally involved miniaturizing
existing technology while striving for improved sensitivity.
   Radioisotope detection is used widely in HPLC but has received
little attention in capillary electrophoresis applications (3-6).
The availability of an on-line radioisotope detector for CE is
especially appealing for several reasons. First, state-of-the-art
radiation detection technology offers extremely high sensitivity.
Second, radioisotope detection affords unrivaled selectivity because
only radiolabeled sample components yield a response at the
detector. Third, the radiolabeled molecule possesses the same
chemical properties as the un-labeled molecule, thereby permitting
tracer studies. Fourth, radioisotope detection can be directly
calibrated to provide a measurement of absolute concentration of the
labeled species. Finally, a capillary electrophoresis system in
which radioactivity detection is coupled with more conventional
detectors adds extra versatility to this new separation technique.
   Radioisotope detection of ^{32}P, ^{14}C, and ^{99}Tc was reported by
Kaniansky et al. (7,8) for isotachophoresis. In their work,
isotachophoretic separations were performed using fluorinated
ethylene-propylene copolymer capillary tubing (300 μm internal
diameter) and either a Geiger-Mueller tube or a plastic
scintillator/photomultiplier tube combination to detect emitted β
particles. One of their reported detection schemes involved passing
the radiolabeled sample components directly through a plastic
scintillator. Detector efficiency for ^{14}C-labeled molecules was
reported to be 13-15%, and a minimum detection limit of 0.44 nCi was
reported for a 212 nL cell volume.
   Altria et al. reported the CE separation and detection of
radiopharmaceuticals containing ^{99m}Tc, a γ emitter with a 6-hour
half-life (9, see also 10). Their design involved passing a
capillary tube ($\approx$ 2 cm long) through a solid block of scintillator
material and detecting the light emitted as technetium-labeled
sample zones traversed the detection volume. Unfortunately,
detection limits and detector efficiency were not reported.

We report here the design and characterization of three
simple, on-line radioisotope detectors for capillary
electrophoresis. The first detector utilizes a commercially
available semiconductor device responding directly to γ rays or β
particles that pass through the walls of the fused silica separation
channel. A similar semiconductor detector for γ-emitting
radiopharmaceuticals separated by HPLC was reported by Needham and
Delaney (11). The second detector utilizes a commercially available
plastic scintillator material that completely surrounds (360°) the
detection region of the separation channel. Light emitted by the
plastic scintillator is collected and focused onto the photocathode
of a cooled photomultiplier tube. Alternatively, a third detection
scheme utilizes a disk fashioned from commercially available plastic
scintillator material positioned between two-room temperature
photomultiplier tubes operated in the coincidence counting mode.
This third scheme maintains favorable collection geometry (360°)
while minimizing detector background noise by electronically
rejecting non-coincident photomultiplier pulses. The detectors
described in the present work are applicable to both high-energy β
emitters and γ emitters. We report here on their application to the
detection of ^{32}P-labeled molecules separated by capillary electrophoresis.

Experimental Section

On-Line Radioactivity Detectors. Our first on-line radioactivity
detector (see Figure 1) consisted of a model S103.1/P4
spectroscopic-grade cadmium telluride semiconductor detector and a
model CTC-4B radiation counting system (Radiation Monitoring
Devices, Inc., Watertown, MA). The cadmium telluride detector probe
consisted of a 2-mm cube of CdTe set in a thermoplastic and
positioned behind a thin film of aluminized mylar at a distance of
approximately 1.5 mm from the face of an aluminum housing (see
Figure 1). A Pb aperture (2 mm wide, 0.008 inch thick) shielded the
CdTe detector element from radiation originating from regions of the
capillary adjacent to the detection volume. The aluminum housing
incorporated a BNC-type connector that facilitated both physical and
electrical connection to a miniature, charge-sensitive preamplifier.
The CdTe probe and preamplifier assembly were mounted on an x-y
translation stage and the face of the aluminum housing was brought
into direct contact with the polyimide-clad fused-silica
capillary/Pb aperture assembly. The CdTe detector was operated at
the manufacturer's suggested bias voltage of 60 V and the detector
signal (the creation of electron-hole pairs produced as β particles
were decelerated within the semiconductor material) was amplified by
the charge-sensitive preamplifier and sent through a six-foot cable
to the counting and display electronics of the CTC-4B counting unit.
Although the CTC-4B is capable of room-temperature energy
discrimination, all experiments reported here were performed with a
relatively large energy window. The upper energy discriminator
setting was 1 MeV (the maximum setting for the CTC-4B) and the low
energy setting was 0.01 MeV.
    The CdTe radioactivity detector was computer interfaced to a
laboratory microcomputer (IBM PC-XT) by using the open collector

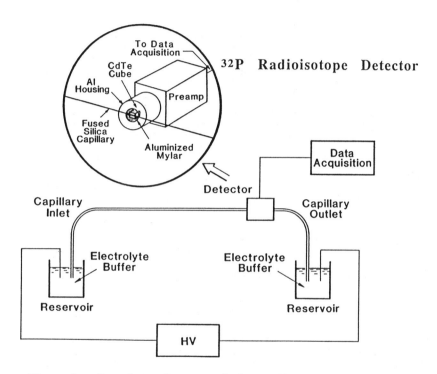

Figure 1. Experimental setup of the capillary
electrophoresis/radioisotope detector system. The inset
shows the positioning of the CdTe probe with respect to the
capillary tubing. The 2-mm Pb aperture is not shown in this
illustration.

output of the CTC-4B counting unit. The open collector output was
tied high by way of a 1-KΩ pull-up resistor so that the unit
provided a negative-going TTL pulse for each count measured. This
TTL signal was sent to a photon counter (model 1109, EG&G Princeton
Applied Research, Princeton, NJ) and counting intervals (typically 1
second) for run-time data acquisition were preselected by way of
front-panel thumbwheel switches on the photon counter. The binary
coded decimal (BCD) output of the photon counter was read at the end
of a preset counting interval (strobe sent by the 1109 counter) by a
laboratory microcomputer (IBM PC-XT) using a 32-bit digital I/O
board (Model DT2817, Data Translation, Inc., Marlboro, MA). Data
acquisition and storage were accomplished using software written in-
house (BASIC). Migration times and peak widths reported here were
determined manually from scale-expanded portions of the recorded
electropherograms.
       Our second on-line radioactivity detector consisted of a
plastic scintillator material (BC-400, Bicron Corp., Newbury, OH)
that was machined from 1-inch-diameter rod stock into a 5/8-inch-
diameter (front face) solid parabola (see Figure 2). A special
rotating holder was constructed for the plastic scintillator and the
curved outer surfaces were coated by vacuum deposition with a thin
film of aluminum in order to reflect the emitted light toward the
front face of the scintillator. A detection length of 2 mm was
defined within the parabola by aluminum mounting rods (0.250 inch
outer diameter) that were press-fit (coaxial to the separation
capillary) in the sides of the scintillator, as illustrated in
Figure 2.
       As radiolabeled sample passed through the detection region,
the scintillator emitted light, which was collected and focused onto
the photocathode of a cooled photomultiplier tube (Centronic # 4283)
by a condenser lens combination (Physitec, # 06-3010, focal length
16 mm). Photon counting was accomplished using a Model 1121A
discriminator control unit and a Model 1109 photon counter (EG&G
Princeton Applied Research). The background count rate observed
under typical operating conditions for this system was approximately
12 counts per second.
       Our third on-line radioactivity detector consisted of a
modified HPLC radioisotope detector (Model 171 Radioisotope
Detector, Beckman Instruments, Inc., Palo Alto, CA). The standard
HPLC flow cell was removed and the unit was modified for use as a
capillary electrophoresis detector, as illustrated in Figure 3. A
2-mm-wide disk of the plastic scintillator material surrounded the
detection region of the separation capillary and was positioned
between two photomultiplier tubes operated in the coincidence
counting mode. For all experiments reported here, the coincidence
gate was 20 nanoseconds. In this manner, most of the random
background pulses associated with the two room temperature
photomultiplier tubes were rejected. A large number of photons were
emitted isotropically for each β particle decelerated within the
scintillator material and a burst of light was thereby sensed at
both of the photomultiplier tubes within the gated time interval.
The background count rate of the modified coincidence detector used
in this work was approximately 30 counts per minute. Data from this
detector was acquired and analyzed using the Chromatographics

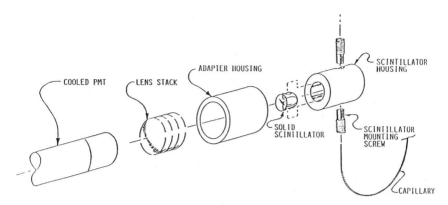

Figure 2. Exploded diagram of the plastic scintillator radioisotope detector. The fused silica capillary is exposed to a 2-mm section of the plastic scintillator located between the press-fit aluminum mounting rods.

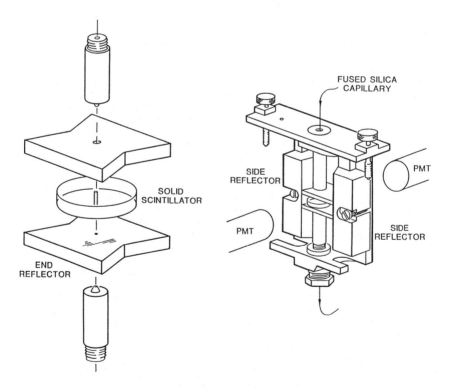

Figure 3. Exploded diagram showing the design of the coincidence radioisotope detection scheme. The fused-silica capillary is exposed to a 2-mm length of plastic scintillator material located between two photomultiplier tubes operated in the coincidence counting mode.

software package (Beckman Instruments, Inc.). Both of the optical
detection systems yielded a large response (at the onset and
completion of each separation), which was associated with the
application or termination of high voltage. The magnitude of this
response was dependent upon the length of time that the system was
held at zero potential between runs, with longer zero potential
periods corresponding to larger responses at the onset of voltage.
Although the cause of this response is not yet fully understood,
the signal was observed to damp out within the first 1-2 minutes of
a run, and therefore posed no interference to the separations
reported here.

Apparatus. The experimental setup of the home-built capillary
electrophoresis system was similar to that described previously
(12,13) and is illustrated in Figure 1. For the semiconductor
detector, a 2-mm section of a fused-silica capillary tube (100 μm
internal diameter, 365 μm outer diameter, 100 cm long, TSP 100/365,
Polymicro Technologies, Inc.) was exposed to the CdTe semiconductor
by placing the Pb foil aperture directly between the face of the
detector housing and the fused silica capillary at a distance of 75
cm from the inlet end of the capillary tube. For the parabolic
plastic scintillator detector, a 2-mm section of a fused-silica
capillary (100 μm internal diameter, 140 cm long) was exposed to the
scintillator by passing the capillary through a 400-μm hole drilled
through the aluminum mounting rods and the central 2-mm portion of
the parabola at a distance of 75 cm from the capillary inlet. For
the coincidence detection scheme a 2-mm section of fused-silica
capillary (100 μm internal diameter, 100 cm long) was exposed to the
plastic scintillator disk by passing the capillary through a 400-μm
hole drilled through both the aluminum shields and the disk itself.
The length of capillary extending from the inlet to the detection
region was 55 cm in the efficiency experiments reported here. This
resulted in a detection volume of approximately 15 nL for each of
the three detectors.

    Each end of the capillary tubing was dipped into a 4-mL glass
vial containing approximately 3 mL of electrolyte-buffer solution.
A strip of Pt-foil submersed in each of the buffer reservoirs
provided connection to high voltage. Plexiglass shielding (0.25 in
thick) was placed around the inlet buffer reservoir because the top
of this vial was quickly contaminated by sample solution carried on
the outside of the capillary tube during the sample injection
procedure. This contamination, if unshielded, lead to unnecessary
operator exposure to radiation.

    The current through the system was monitored as a potential
drop across a 1-KΩ resistor in the circuit. The capillary system
and detector were enclosed in a Plexiglass box to prevent operator
exposure to high voltage. Electroosmotic flow rates for free-
solution separations reported here were measured in a manner similar
to that described by Huang et al. (14). The capillary tube was
filled with running buffer diluted by 10%, the buffer reservoirs
were filled with running buffer, and the current was monitored as
one tubing volume was displaced by supporting buffer under the
influence of the applied potential. Sample introduction in all
free-solution separations reported here was accomplished by

hydrostatic pressure. Electrokinetic sample introduction was used
for all capillary gel separations.

The high-voltage power supply (Model MG30N100, Glassmann High
Voltage, Whitehouse Station, NJ) was continuously programmable from
0 to -30 kV by means of an external 0-to-10-volt DC signal. The
flow-programming experiments reported here were accomplished by
manually reducing the program voltage to the high-voltage supply.

Reagents. Aqueous ethanol solutions of the triethylammonium salts
of adenosine-5'-[α-^{32}P]triphosphate (α-^{32}P ATP), adenosine-5'-[γ-
^{32}P]triphosphate (γ-^{32}P-ATP), thymidine-5'-[α-^{32}P]triphosphate (α-
^{32}P-TTP), cytidine-5'-[α-^{32}P]triphosphate (α-^{32}P-CTP), and
guanosine-5'-[α-^{32}P]triphosphate (α-^{32}P-GTP) were purchased from
Amersham (Arlington Heights, IL). Radioactive sample concentrations
reported for detector efficiency determination were adjusted from
the manufacturer's specifications after subjecting several diluted
aliquots of the stock solution to liquid scintillation counting.
The concentration was further corrected for radiochemical purity
according to the manufacturer's specifications because liquid
scintillation counting measures the total sample activity and does
not account for the presence of radiolabeled impurities. Stock
solutions were stored at -15°C or -20° C in order to minimize sample
loss due to hydrolysis. Injected sample solutions were prepared in
0.25 mL plastic vials by diluting stock solutions with buffer or
deionized water and were also stored at -15° C or -20° C.

The supporting electrolyte for all free-solution separations
reported here was a borate buffer (pH 8.1 or 8.26, 0.10 M or 0.20
M), prepared from reagent-grade sodium borate decahydrate and boric
acid (J.T Baker).

For the preparation of gel-filled capillaries, a solution
consisting of 50 mM tris, 50 mM boric acid, and 7 M urea (pH = 8.3)
was used both to prepare the gel and as the running buffer. For
some of the gel-filled capillaries used in this work, the buffer
also contained 3% PEG 20,000 (15) (Fluka Chemical Corp., Ronkonkoma,
NY). The fused-silica capillaries were rinsed with 1 N HCl, 1N
NaOH, and then methanol. A 1:1 mixture of
methacryloxypropyltrimethoxysilane and methanol was introduced into
the capillary (100 or 75 μm internal diameter) by syringe and
allowed to sit at room temperature for at least 3 hours. The
acrylamide/crosslinker (N, N' methylenebisacrylamide) solution was
prepared in 10 ml of running buffer to yield a gel composition of
either 7% T, 3% C, or 6% T, 5% C. To initiate polymerization, 2-5
μL aliquots of 10% solutions of both ammonium persulfate (APS) and
N,N,N',N'-tetramethylethylenediamine (TEMED) were added to 1 mL of
the monomer solution. The monomer solution was then quickly
introduced into the capillary by syringe and left at room
temperature overnight for complete polymerization. Once
polymerization was complete, care was taken to keep the capillary
ends submerged in buffer at all times in order to prevent the gel
from drying out.

The oligonucleotide samples described here were labeled at the
5' end according to the following procedure. Approximately 0.5 OD
of oligomer sample was combined with 5 μL of γ-^{32}P ATP (10 mCi/ml,
5000 ci/mmole, Amersham), 5 μL of 10x kinase reaction buffer

(Amersham), and 2 μL of kinase. The total solution volume was
adjusted to 50 μL by the addition of HPLC grade water. The solution
was gently mixed and allowed to react at 37° C for one hour. At the
end of this period, the labeled oligomer sample was isolated from
unincorporated γ-^{32}P ATP by either ethanol precipitation or by size
exclusion chromatography using Sephadex G-25. The labeled oligomer
was then lyophilized and dissolved in an appropriate volume of
deionized water before being electrokinetically injected into the
gel-filled capillaries.

Results and Discussion

The process of β decay for ^{32}P can be written as (16),

$$\underset{15}{\overset{32}{P}} \rightarrow \underset{16}{\overset{32}{S}} + \beta^- + \bar{v} \tag{1}$$

where β^- represents the negatively charged β particle and $\bar{v}$ is the
antineutrino. ^{32}P is an example of a "pure β emitter" that
populates only the ground state of the product nucleus. Each β-
decay transition is characterized by a fixed decay energy shared
between the β particle and the antineutrino. As a result, the β
particle is emitted with an energy that varies from decay to decay
and ranges from zero to the "β end-point energy," which is
numerically equal to the transition decay energy. A β-energy
spectrum for ^{32}P shows a maximum particle energy of 1.7 MeV and an
average particle energy of approximately 0.57 MeV. The penetrating
ability of β particles through various media may be obtained from
literature range-energy plots in which the product of particle range
and medium density ("mass thickness") is plotted against particle
energy. Such plots are especially useful because they may be used
to predict the penetration length at a given particle energy in
media other than that used to obtain the original plot (16). From
such plots, one would predict that the average β particle energy ($\approx$
0.57 MeV) produced by decay of ^{32}P would correspond to a range of
approximately 2000 μm in water and approximately 950 μm in fused
silica. Thus, ^{32}P decay would be detectable by a sensitive device
positioned external to the fused silica capillary tubing (of the
dimensions normally selected for capillary electrophoresis
separations).
        Successful detection of ^{32}P-labeled molecules separated by
capillary electrophoresis using the above detection schemes, in
which a sensor was positioned external to the separation channel,
was made possible by several factors. These included (1) the large
energy associated with β decay of ^{32}P (1.7 MeV), (2) the high
sensitivity and small size of commercially available semiconductor
detectors, (3) the availability of efficient solid scintillator
materials and sensitive photomultiplier tubes, (4) the short lengths
of fused silica (capillary wall thickness) and aqueous electrolyte
through which the radiation must pass before striking the detector,
and (5) the relatively short half-life of ^{32}P (14.3 days).
        Because the CdTe detector was not visible through the
aluminized mylar film, it was necessary to check for proper

alignment of the capillary tube with respect to the CdTe cube. This
was accomplished by filling the detection volume with radioactive
material and monitoring the signal level as the detector was
translated with respect to the capillary. The observed signal was
not very sensitive to positioning when the capillary was offset over
a range of ± 1.5 mm from the center of the aluminum housing but
dropped off rapidly at greater distances. All experiments reported
here were performed with the capillary positioned at the center of
the aluminum housing, as indicated in Figure 1.

Detector Efficiency. In order to calculate the efficiency of the
on-line radioactivity detectors for ^{32}P, it was necessary to
determine the volume of sample injected onto the capillary tube by
the gravity-flow technique. The volume of sample introduced by
hydrostatic pressure was determined as follows: A plug of ^{32}P-
labeled ATP was introduced onto the capillary by raising the sample
vial above the high-voltage reservoir for a carefully timed
interval. The end of the capillary was then returned to the anode
reservoir and electrophoresis was performed for 5 minutes at high
voltage. This 5-minute high-voltage period served to transfer the
sample plug toward the detector and away from the injection end of
the capillary as if an actual separation were being performed. At
the end of the 5-minute period the voltage was switched off and the
electrolyte within the capillary tube was driven, via syringe, into
a liquid scintillation vial located beneath the capillary outlet.
This process was continued until approximately 8 capillary volumes
of electrolyte were collected. The collected sample plugs were
mixed with scintillation cocktail and subjected to liquid
scintillation counting. The injection volumes were determined by
relating the activity of the sample plugs to that of the injected
sample solution. The injection volumes and repeatability of the
manually performed hydrostatic injections for the three detector
efficiency determinations are shown in Tables I, II, and III.

Table I.   Injection Data for CdTe Radioisotope
Detector CE System

Injection No.	DPM
1	101885
2	110449
3	101884
4	111375
5	103512
6	113018
7	109432
8	104581
9	106740
10	107357
Average	107023
Std. Dev.	3996
% RSD	3.7
Injection Volume	84 nL

Table II. Injection Data for Plastic Scintillator
Radioisotope Detector CE System

Injection No.	DPM
1	89732
2	95220
3	90353
4	94939
5	90564
6	90039
7	94648
8	96628
9	95212
10	93150
Average	93049
Std. Dev.	2621
% RSD	2.8
Injection Volume	72 nL

Table III. Injection Data for Coincidence
Radioisotope Detector CE System

Injection No.	DPM
1	37167
2	38220
3	40071
4	39010
5	32237
6	36784
7	35845
8	36032
9	35606
10	36767
Average	36773
Std. Dev.	2146
% RSD	5.8
Injection Volume	60 nL

Replicate capillary electrophoresis runs were made in which a
standard solution of ^{32}P-labeled ATP was injected into the
capillary. The results are shown in Tables IV, V, and VI. These
tables list the migration time, peak area, residence time, and
detector efficiency. Representative electropherograms corresponding
to the three detector efficiency determinations and the conditions
under which the separations were performed are shown in Figures 4,
5, and 6. The efficiencies reported in the tables were calculated
using the following equation:

$$NOC = (DPM_{peak})(Residence\ Time)(Efficiency)$$

$$= (DPM_{peak})(Detector\ Length/Zone\ Velocity)(Efficiency), \quad (2)$$

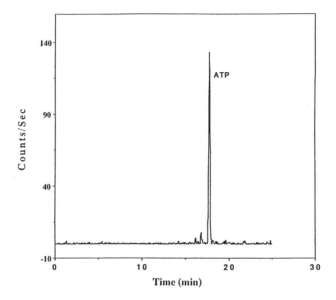

Figure 4.   Capillary electropherogram of adenosine-5'-[α-
^{32}P] triphosphate obtained by injecting approximately 51 nCi
(7 x 10^{-8} M solution) onto the capillary and applying a
constant potential of -20 kV.   The separation was
continuously monitored using the CdTe semiconductor
radioisotope detector.   Data were subjected to a 5-point
sliding smooth.   Electrolyte was 0.2 M borate buffer, pH =
8.1.

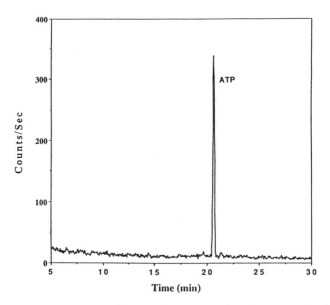

Figure 5. Capillary electropherogram of adenosine-5'-[α-
^{32}P] triphosphate obtained by injecting approximately 38 nCi
(6 x 10^{-8} M solution) onto the capillary and applying a
constant potential of -25 kV. The separation was monitored
using the parabolic plastic scintillator radioisotope
detector. Data were subjected to a 5-point sliding smooth.
The electrolyte was the same as in Figure 4.

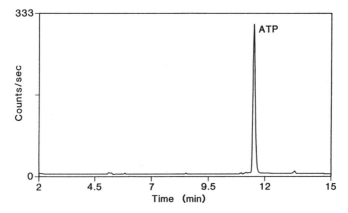

Figure 6. Capillary electropherogram of adenosine-5'-[γ-
^{32}P] triphosphate obtained by injecting approximately 25 nCi
(6 x 10^{-8} M solution) onto the capillary and applying a
constant potential of -25 kV. The separation was monitored
using the coincidence radioisotope detector.

where "NOC" represents the number of observed counts integrated over a peak, "DPM$_{peak}$" represents the number of radioactive transformations occurring each minute in the injected sample plug, "Residence Time" is the amount of time (in minutes) a radioactive molecule within a given sample zone spends in the detection volume, and "Efficiency" is the fractional number of events sensed by the detector. The efficiencies for the on-line detectors described here

Table IV. CdTe Radioisotope Detector Efficiency Data

Run	Elution Time, mins	Peak Area, counts	Residence Time, mins	Efficiency
1	18.33	1519	0.049	27.6%
2	17.96	1372	0.048	25.4%
3	18.30	1461	0.049	26.5%
4	18.07	1511	0.048	28.0%
5	17.98	1407	0.048	26.1%
6	17.48	1601	0.047	30.3%
7	17.76	1197	0.047	22.7%
8	17.32	1195	0.046	23.1%
9	17.91	1392	0.048	25.8%
10	17.37	1266	0.046	24.5%
11	17.77	1359	0.047	25.7%
av	17.84	1389		26.0%
sd	0.34	132		2.2%
%RSD	1.9	9.5		8.4%

Table V. Parabolic Radioisotope Detector Efficiency Data

Run	Elution Time, mins	Peak Area, counts	Residence Time, mins	Efficiency
1	20.65	3375	0.055	72.7%
2	20.62	3088	0.055	66.5%
3	20.55	3119	0.055	67.2%
4	20.54	3091	0.055	66.5%
5	20.69	2953	0.055	63.6%
6	20.84	2881	0.056	60.9%
7	21.22	2883	0.057	59.9%
8	21.00	2952	0.056	62.4%
9	20.84	3213	0.056	67.9%
10	20.88	3122	0.056	66.0%
av	20.78	3068		65.4%
sd	0.22	156		3.8%
%RSD	1.06	5.07		5.8%

are largely a function of detector collection geometry, *i.e.*, positioning of the CdTe probe or plastic scintillator with respect to the capillary. Note that the residence time within the detector must be determined for each component in a mixture because separated sample zones travel with different velocities according to their individual electrophoretic mobilities. This is in sharp contrast to radio-HPLC detection, in which the residence time for each sample component is the same and is given by the ratio of the detector cell

volume to the mobile phase flow rate. The residence time for a
particular sample component separated by capillary electrophoresis
is easily obtained from its migration time and from the length of
capillary to which the detector is exposed.

Table VI. Coincidence Radioisotope Detector Efficiency Data

Run	Elution Time, mins	Peak Area, counts	Residence Time, mins	Efficiency
1	11.52	2236	0.042	103%
2	11.58	2196	0.042	101%
3	11.87	2031	0.043	92%
4	11.65	2277	0.042	105%
5	11.58	2237	0.042	103%
6	11.45	1966	0.042	91%
7	11.18	1936	0.041	91%
8	10.75	2038	0.039	101%
9	10.92	2182	0.040	112%
10	11.05	2311	0.040	101%
11	11.77	2072	0.043	93%
12	11.02	2200	0.040	106%
av				100%
sd				6.97
%RSD				6.97%

Results obtained for the replicate runs shown in Tables IV, V,
and VI indicate that the measured efficiency of ^{32}P detection for
the on-line CdTe radioactivity detector is approximately 26%, while
the efficiency of the plastic scintillator radioactivity detectors
is approximately 65% (parabola) and 100% (coincidence unit),
reflecting the improved geometry of the latter two devices. The
background noise level of the CdTe detector system is a function of
the low energy discriminator setting. The value of 0.01 MeV
selected for all experiments reported here gave a background count
rate of approximately 10 counts per minute while leaving a wide
energy window open for detection. Comparison of signal-to-noise
ratios in the three electropherograms indicates that the three
detectors exhibit quite similar sensitivities despite the fact that
the efficiency of the plastic scintillator detectors is considerably
greater than that of the semiconductor detector. This difference in
sensitivity is caused by the extremely low background noise level of
the CdTe device compared with a photomultiplier tube.
     The large gain in sensitivity afforded by on-line radioisotope
detection in comparison with the more commonly used UV-absorbance
detector is illustrated in Figure 7. In this example, a UV-
absorbance detector, monitoring at 254 nm, was positioned 8.5 cm
downstream from a CdTe radioisotope detector, and ^{32}P-labeled ATP
was injected at a concentration of approximately 5×10^{-8} M. Under
these conditions, ATP is detected with an excellent signal-to-noise
ratio by the radioisotope detector but is completely undetectable by
UV absorbance.

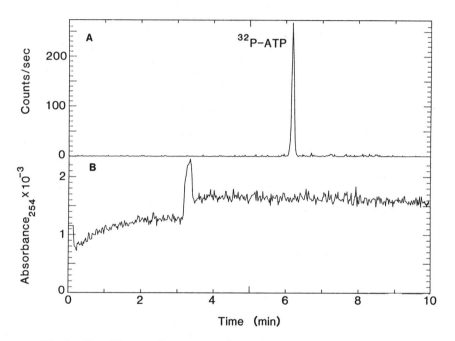

Figure 7. Electropherograms showing (A) CdTe radioisotope detector response and (B) UV absorbance detector response. The injected sample was 5×10^{-8} M ^{32}P-labeled ATP. The UV absorbance detector was located 8 cm downstream from the radioisotope detector, and absorbance was monitored at 254 nm.

Flow Programming. Equation 2 suggests that the number of counts
measured (the detector sensitivity) over a sample peak may be
increased by lengthening the residence time of the sample in the
detection volume. This is equivalent to increasing the counting
time on a liquid scintillation counter and this concept has been
recognized in radio-HPLC applications (17) and applied to
radioisotope detection in isotachophoresis (7). In capillary
electrophoresis, the velocity of a sample zone may be reduced and
its residence time increased by simply reducing the applied
potential as the zone passes through the detection volume. The most
efficient implementation of this flow-programming concept would
involve reducing the zone velocities only while the labeled sample
was present within the detection volume and operating at a
relatively high potential at all other times. To our knowledge,
this type of flow programming has not previously been explored in
capillary electrophoresis. Although it is demonstrated only for
radioisotope detection here, this methodology should be applicable
to other modes of sample detection in CE.
        The flow-programming concept is demonstrated in Table VII,
which lists the peak width and peak area for six capillary

Table VII. Flow-Programmed Runs

Run	Elution Time, mins	Peak Width, mins	Peak Area, counts	Voltage Program
1	18.00	0.38	985	20 kv constant
2	18.34	0.34	1098	
3	19.36	0.43	1236	
4	18.04	0.38	1016	
5	18.75	0.48	1078	
6	18.14	0.35	968	
Average =		0.39	1064	
7	18.75	0.87	2404	20 kv initial
8	18.45	0.79	2705	10 kv during
9	17.30	0.71	2081	detection
10	17.65	0.81	2448	period
11	16.82	1.12	2695	
12	19.02	0.80	2673	
Average =		0.85	2501	
Peak Area Ratio	2.4			
Current Ratio	2.4			
Voltage Ratio	2.0			

electrophoresis separations performed with and without flow
programming. Separations 1 through 6 were performed at a constant
potential of -20 kV using the CdTe radioisotope detector, while in
runs 7 through 12 the potential was reduced to -10 kV as soon as
signal was detected above the detector background level. Because
the zone velocity is directly proportional to the applied field
strength, the average temporal peak width and area (number of counts
observed) for the six flow-programmed runs were approximately

doubled. This improvement in sensitivity is, however, accompanied
by an increase in analysis time as well as a small loss in
resolution due to zone broadening. The magnitude of the resolution
losses incurred during flow programming will be strongly dependent
upon the amount of sample injected and the additional run time
associated with the flow programming process. For injected sample
plug lengths several times larger than the length associated with
diffusional broadening (typical operating conditions), the
resolution loss will not be significant. In the limit of injected
sample plugs with no initial width (δ function), the additional peak
variance increases linearly with programming time (ignoring analyte-
wall interactions) and the resolution loss will become quite
significant.

A striking example of increased sensitivity gained through the
application of flow programming is illustrated in Figures 8 and 9.
In Figure 8, a synthetic mixture of thymidine-5'-[α-^{32}P]triphosphate
(^{32}P-TTP), cytidine-5'-[α-^{32}P]triphosphate (^{32}P-CTP), and adenosine-
5'-[α-^{32}P]triphosphate (^{32}P-ATP), with each component present at a
concentration of approximately 3×10^{-8} M (≈ 19 nCi injected), was
injected using hydrostatic pressure and separated under the
influence of a constant -20 kV applied potential. In Figure 9, the
sample solution and injection volume were the same as in Figure 8,
but the residence time of each component was increased by reducing
the applied potential from -20 kV to -2 kV as radioactive sample was
passing through the detection volume. At the same time, the
counting interval was increased proportionately (from 1 second to 10
seconds) and the detector signal was plotted as a function of
electrolyte volume displaced through the capillary tube. Note that
this results in a time-compressed abscissa over the flow programmed
period of the electropherogram (the entire separation required about
70 minutes in this case). It is important to point out that the
lower limit of radioisotope detection refers to the lowest sample
activity contained within a peak that can be detected and accurately
quantified. From the data presented in Table VII and Figures 4-9 it
is apparent that the lower limit of detection for this system is
greatly dependent upon the conditions under which the analysis is
performed, and that detector sensitivity may be extended by an order
of magnitude or more using flow programming.

The sensitivity gain afforded by this flow-programming
methodology will ultimately be limited by practical considerations
of analysis time and resolution losses caused by diffusional
broadening of the sample zones. Simplicity and consideration of
analysis time, however, still make flow-counting detection for
capillary electrophoresis an attractive alternative to the
quantitatively superior batch-counting approach in which fractions
are collected and subjected to conventional counting techniques
(17). The batch-counting approach, provided that sufficiently small
fractions may be collected, does offer the advantage of decoupling
separation considerations from measurement time. Considering only
the limitation imposed by diffusional spreading of sample zones
during the flow-programmed portion of a run, it is possible to
predict the extent to which detector sensitivity may be improved by
flow programming. For an injection volume of 84 nL, as used in this
example, and a maximum allowable increase in zone variance defined

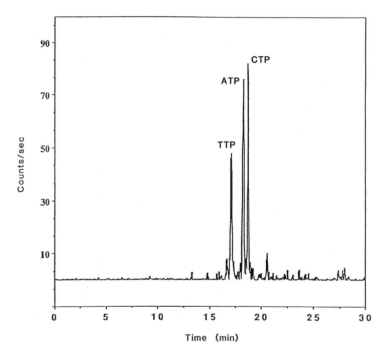

Figure 8. Capillary electropherogram of thymidine-5'-[α-
^{32}P] triphosphate, cytidine-5'-[α-^{32}P] triphosphate, and
adenosine-5'-[α-^{32}P] triphosphate obtained by injecting
approximately 19 nCi (2 x 10^{-8} M solution) of each component
onto the capillary and applying a constant potential of -20
kV.

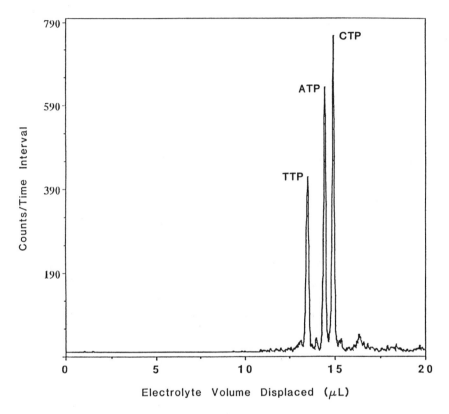

Figure 9. Flow-programmed capillary electropherogram of
thymidine-5'-[α-^{32}P] triphosphate, cytidine-5'-[α-^{32}P]
triphosphate, and adenosine-5'-[α-^{32}P] triphosphate obtained
by injecting approximately 19 nCi (2 x 10^{-8} M solution) of
each component onto the capillary. The separation was flow
programmed by applying a constant potential of -20 kV until
radiolabeled sample approached the detection volume and then
reducing the potential to -2 kV as the sample zones
traversed the detection region. Note that the detector
signal is plotted as a function of electrolyte volume
displaced, resulting in a time-compressed abscissa over the
flow programmed region of the electropherogram. The
operating current was 38 μA at -20 kV and 3.8 μA at -2 kV.
The data were subjected to a 5-point sliding smooth.

to be 10%, approximately 84 minutes of flow programming would be
permitted (this calculation assumes a rectangular injection profile
and a solute diffusion coefficient of 10^{-6} cm^2/sec, and neglects
both diffusional broadening prior to flow programming and velocity-
dependent analyte-wall interactions). This 10% increase in variance
would be accompanied by a 250-fold increase in the number of counts
observed over a peak. Because the sensitivity of radioisotope
detection is governed by counting statistics, a 16-fold increase in
the signal-to-noise ratio (NOC/(NOC)$^{-\frac{1}{2}}$) would result. Thus, a lower
limit of detection of about 10^{-11} M would seem to be a conservative
extrapolation. Obviously, the limitations imposed by diffusional
broadening would become more severe if the initial injection volume
were reduced.

In an automated implementation of the flow-programming
methodology, that is, with the high-voltage power supply under
computer control, there is a further limitation imposed upon
achievable sensitivity gains. There must be enough sample present
to generate a signal sufficiently large to exceed the detector
background level under standard (non-flow-programmed) conditions, in
order to initiate the flow-programming procedure. In certain
instances, however, prior knowledge of elution times for the
compounds of interest would permit this limitation to be overcome.

Application to Capillary Gel Electrophoresis. Recently, Karger and
co-workers demonstrated the use of polyacrylamide gel-filled
capillaries to separate peptide/protein (SDS PAGE) (18) and
oligonucleotide mixtures (19,20) by capillary electrophoresis. This
mode of CE operation may prove to couple well with on-line
radioisotope detection. The results of several preliminary
capillary electrophoresis separations using gel-filled capillaries
and on-line radioisotope detection using the coincidence unit
described here are presented below.

The capillary gel electrophoresis separation of a three-
component, ^{32}P-labeled, nucleotide mixture is illustrated in Figure
10. It is interesting to note that the migration order of ATP and
CTP is reversed with respect to the free solution separations
presented in Figures 8 and 9. This is caused by the absence of
electroosmotic flow in the gel-filled capillary.

Figure 11 illustrates the CE separation of synthetic
polythymidylic oligomers. The capillary gel electrophoresis
separation of this sample has previously been described by Paulus
and Ohms (21) using UV-absorbance detection. The polythymidylic 50-
mer sample was synthesized with the reaction conditions purposely
adjusted to increase the failure rate at every fifth base, beginning
with the 15-mer.

Figures 12 and 13 illustrate the CE separation of ^{32}P-labeled
29- and 30-base heterooligomers, respectively. In the two
electropherograms, the major component is nicely resolved from
several failure sequences that were also phosphorylated in the
labeling procedure of the 5' end. Figure 14 illustrates the
capillary gel separation of a mixture containing these two
heteropolymers. These two polymers differ only by the absence or
presence of a 3' terminal thymidine residue.

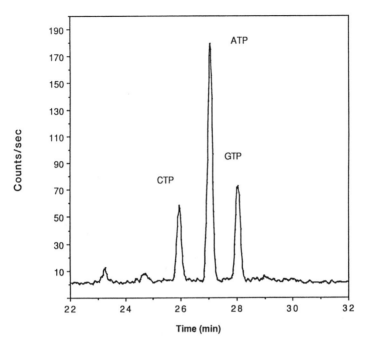

Figure 10. Capillary gel electrophoresis separation of a simple three-component nucleotide mixture with on-line radioisotope detection using the coincidence unit.

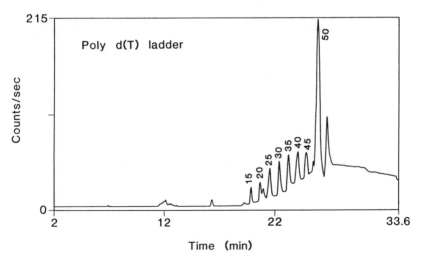

Figure 11. Capillary gel electrophoresis separation a poly (T) oligomer sample ^{32}P-labeled at the 5' end. Detection was accomplished using the coincidence radioisotope detector.

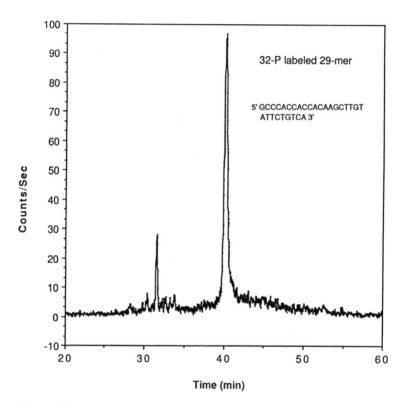

Figure 12. Capillary gel electrophoresis separation of
heteropolymer ^{32}P-labeled at the 5' end. The oligomer was 29
units in length with the sequence shown in the Figure.
Detection was accomplished using the coincidence
radioisotope detector.

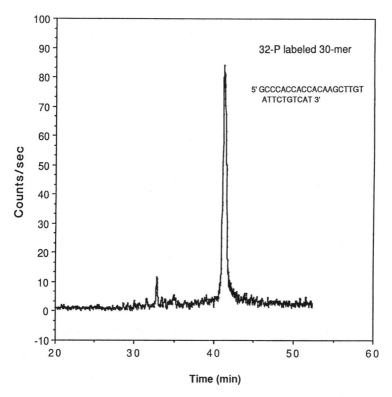

Figure 13. Capillary gel electrophoresis separation of heteropolymer [32]-labeled at the 5' end. The oligomer was 30 units in length with the sequence shown in the figure. Detection was accomplished using the coincidence radioisotope detector.

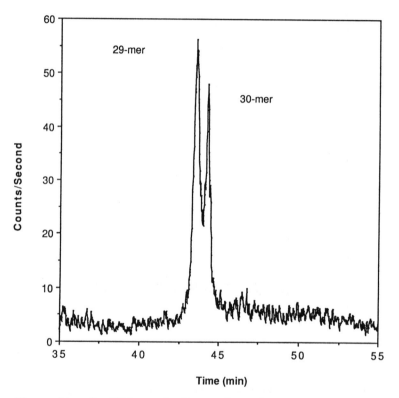

Figure 14.   Capillary gel electrophoresis separation of 29-
and 30-mer, ^{32}P-labeled at the 5' end.   Detection was
accomplished using the coincidence radioisotope detector.

It is worth noting that the sensitivity advantage of radioisotope detection in comparison with UV absorbance detection is not so large for the biopolymers studied here as it is for small molecules such as the nucleotide triphosphates. This is because the labeling scheme at the 5' end transfers only one ^{32}P atom per oligomer molecule while the same oligomer, containing many chromophores, absorbs quite strongly. Hence the sensitivity advantage of radioisotope detection will continue to decline as the length of the oligomer increases. Obviously, a labeling scheme that resulted in the transfer of more radiolabels per molecule would increase the sensitivity difference of the two detectors. Perhaps a more significant advantage of radioisotope detection is that the separation medium need not be optically transparent. This opens up a broader range of buffer components and matrix stabilizers from which the separation medium may be formulated.

The application of flow programming to the analysis of samples using gel-filled capillaries also results in a sensitivity gain. This is illustrated in Figures 15 and 16. Figure 15 shows the CE separation of a polydeoxyadenosine 40-60 mer sample using a polyacrylamide-filled capillary under constant voltage. The electropherogram presented in Figure 16 was obtained using the same sample and column but under flow-programmed conditions. The enhancement in sensitivity is marked.

Autoradiography. Of course, residence time can be arbitrarily increased by removing the applied field. However, there is an obvious trade-off between measurement time and resolution loss caused by diffusional broadening. The latter may be significantly reduced by freezing the capillary contents. This allows an autoradiographic view of the separation channel by exposing directly the frozen capillary to x-ray film (Kodak Diagnostic XAR-5, Rochester, NY). Autoradiography is a well-established technique (26). This method is not limited to high energy β emitters; it can be applied to detect radiolabels that have less penetrating radiation, provided that the capillary contents are formulated to scintillate.

Figure 17 shows the autoradiographic detection of the ^{32}P-labeled polydeoxyadenosine homopolymer sample, which is the same sample as shown in Figures 15 and 16. The gain in sensitivity is several orders of magnitude. Indeed, it is even possible to discern ^{32}P-labeled material "trapped" in the gel matrix between peaks. Clearly this method is the method of choice when sensitivity is an issue. The same method can also be applied to free-solution capillary electrophoresis.

Conclusion

Three simple, on-line radioisotope detectors for capillary electrophoresis were described and characterized for the analysis of ^{32}P-labeled analytes. The minimum limit of detection for these systems was shown to be strongly dependent upon the conditions under which the analysis is performed. For standard CE separations performed at a relatively high (constant) voltage, the minimum limit of detection was found to be in the low nanocurie (injected sample

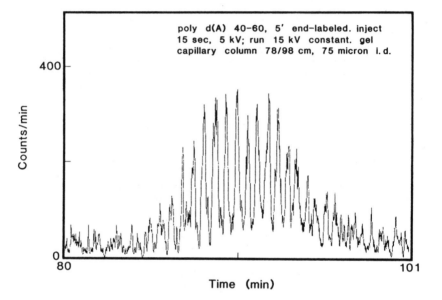

Figure 15. Electropherogram illustrating the capillary
electrophoresis separation of poly d(A) 40-60 mer sample,
[32]P-labeled at the 5' end. Detection was accomplished using
the coincidence detector. The separation was accomplished
using a polyacrylamide gel-filled capillary and a constant
potential of 15 kV. The sample activity in this example was
approximately 4800 DPM/nL.

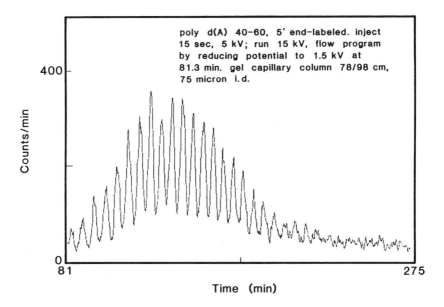

Figure 16. Electropherogram illustrating the flow-
programmed separation of the same ply d(A) 40-60 mer sample
presented in Figure 15. The sample was separated at 15 kV
and the potential was reduced to 1.5 kV as radiolabeled
sample reached the detector. The sensitivity improvement
afforded by flow programming is readily apparent in this
figure.

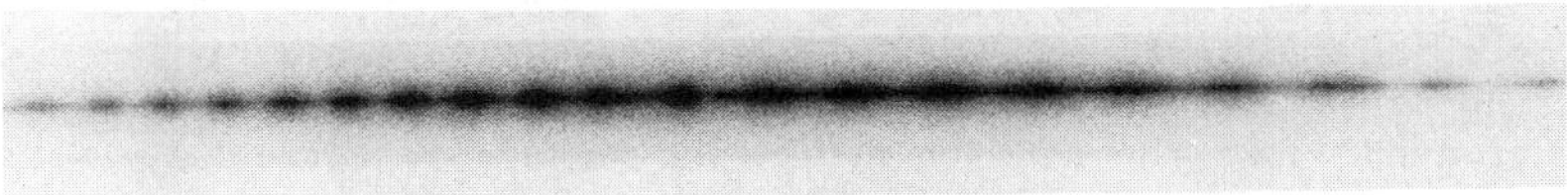

Figure 17. Autoradiogram showing the separated poly d(A)
40-60 mer sample located within a polyacrylamide gel-filled
capillary. The autoradiogram was measured by placing the
capillary on a piece of x-ray film and freezing the
capillary-film combination at -20° C for 15 hours. The
leading end of the sample is on the right in this
photograph.

quantity) range, corresponding to an analyte concentration of about
10^{-9} M. The lower limit of detection for this type of detection
system was extended to the sub-nanocurie level ($\approx 10^{-10}$ M) by
application of flow programming methodology which served to increase
the residence time of labeled sample components within the detection
volume. Further large gains have been demonstrated by freezing the
contents of the capillary after separation and exposing the frozen
capillary to film (autoradiography). Thus radioisotope detection,
when applicable, has a sensitivity superior to most other detection
schemes, comparable to electrochemical detection (22,23) and laser-
induced-fluorescence detection (12,24,25).
    One improvement to the current systems would involve
automation of the flow-programming methodology, and such efforts are
currently underway. A second improvement over the current
semiconductor system would involve optimizing the detector geometry
by capturing a larger solid angle with the CdTe detector. The
performance of the parabolic plastic scintillator detector would be
greatly improved by reducing the background noise level through the
use of a quieter photomultiplier tube. In certain instances it
would be desirable to reduce the effective detection volume of these
systems in order to increase resolution. This could be accomplished
by installing a narrower aperture in the semiconductor detector or
machining a smaller detection region from the plastic scintillator
materials. In either case, detector sensitivity would be reduced,
because the detection volume and effective residence time would be
decreased. Hence, there is once again a practical trade-off between
detector sensitivity and resolution.
    On-line radioisotope detection has been demonstrated to be a
practical alternative to UV absorbance detection when gel-filled
capillaries are used for CE separations. Significant improvement in
detection limits is realized with radioisotope detection. The
greatest improvement is realized for small molecules and is roughly
one to two orders of magnitude (for runs in which the residence time
is not enhanced).
    Future work in this area will focus on the extension of this
detection scheme to include other radioactive isotopes. The present
systems are applicable to high-energy β and γ emitters. Peptide and
protein samples labeled with ^{125}I or ^{131}I should prove to be an
interesting application. Weaker β emitters will require that the
sensing device be placed in direct contact with the electrolyte
solution and that the sensing device be compatible with changing
field gradients.

Acknowledgments

S.L.P. wishes to thank David J. Rakestraw, Patrick H. Vaccaro,
W. Howard Whitted and James Burns for many helpful conversations
pertaining to this work. The assistance of Aaron Paulus, Andras
Guttman, and Sushma Rampal in the preparation of gel-filled
capillaries and Karen Wert in the efficiency determinations is also
gratefully acknowledged.

Credit

Support for this work by Beckman Instruments, Inc. is gratefully
acknowledged.

Literature Cited
1. Nickerson, B.; Jorgenson, J. W. *J. High Resolut. Chromatogr. Chromatogr. Commun.* 1988, *11(7)*, 533-534.
2. Jorgenson, J. W.; Lukacs, K. D. *Science* 1983, *222*, 266-272.
3. Kessler, M. J. In *Analytical and Chromatographic Techniques in Radiopharmaceutical Chemistry*; Wieland, D. M.; Tobes, M. C.; Mangner, T. J., Eds.; Springer-Verlag: New York, 1987; Ch. 5-7.
4. Roberts, T. R. *Radiochromatography*, Journal of Chromatography Library; Elsevier: Amsterdam, 1978, Ch. 6.
5. Kessler, M. J. *Am. Lab.* 1988, *20(6)*, 86-95.
6. Kessler, M. J. *Am. Lab.* 1988, *20(8)*, 76-81.
7. Kaniansky, D.; Rajec, P.; Švec, A.; Havaši, P.; Macášek, F. *J. Chromatogr.* 1983, *258*, 238-243.
8. Kaniansky, D.; Rajec, P.; Švec, A.; Marák, J.; Koval, M.; Lúcka, M.; Franko, Š.; Sabanoš, G. *J. Radioanal. Nucl. Chem.* 1989, *129(2)*, 305-325.
9. Altria, K.D.; Simpson, C.F.; Bharij, A.; Theobald, A.E. Paper presented at the 1988 Pittsburgh Conference and Exposition, abstract no. 642, New Orleans, February 1988.
10. Berry, V. *LC/GC* 1988, *6*, 484-491.
11. Needham, R.E.; Delaney, M.F. *Anal. Chem.* 1983, *55*, 148-150.
12. Gassmann, E.; Kuo, J.E.; Zare, R.N. *Science* 1985, *230*, 813-814.
13. Gordon, M. J.; Huang, X.; Pentoney, S.L., Jr.; Zare, R.N. *Science* 1988, *242*, 224-228.
14. Huang, X.; Gordon, M.J.; Zare, R.N. *Anal. Chem.* 1988, *60*, 1837-1838.
15. Karger, B. L.; Cohen, A. U. S. Patent #4865707, 1989.
16. Knoll, G.F. *Radiation Detection and Measurement*; Wiley: New York, 1979.
17. See, for example, Markl, P. In *Instrumentation for High Performance Liquid Chromatography*; Journal of Chromatography Library, Volume 13; Elsevier: Amsterdam, 1978, pp. 151-161.
18. Cohen, A. S.; Karger, B. L. *J. Chromatogr.* 1987, *397*, 409-417.
19. Guttman, A.; Paulus, A.; Cohen, A. S.; Karger, B. L. Electrophoresis '88, Proc. Int. Electrophoresis Society Meeting, 6th, Copenhagen, 1988, pp. 151-159.
20. Cohen, A. S.; Najarian, D. R.; Paulus, A.; Guttman, A.; Smith, J. A.; Karger, B. L. *Proc. Natl. Acad. Sci.* 1988, *85*, 9660-9663.
21. Paulus, A.; Ohms, J. J. *J. Chromatogr.* 1989, accepted for publication.
22. Wallingford, R. A.; Ewing, A. G. *Anal. Chem.* 1988, *60*, 258-263.
23. Wallingford, R. A.; Ewing, A. G. *Anal. Chem.* 1987, *59*, 1762-1766.
24. Gozel, P.; Gassmann, E.; Michelsen, H.; Zare, R. N. *Anal. Chem.* 1987, *59*, 44-49.
25. Dovichi, N. J. Paper presented at the 41st ACS Summer Symposium on Analytical Chemistry, Stanford University, 26 to 29 June 1988.
26. Stryer, L. *Biochemistry*; Freeman and Company, 1981, Ch. 24.

RECEIVED December 20, 1989

Chapter 5

Strategies for an Analytical Examination of Biological Pharmaceuticals

Eleanor Canova-Davis, Glen M. Teshima, T. Jeremey Kessler, Paul-Jane Lee, Andrew W. Guzzetta, and William S. Hancock

Genetech, Inc., 460 Point San Bruno Boulevard, South San Francisco, CA 94080

The chemical complexity of protein pharmaceuticals has resulted in the requirement of a battery of analytical methods for product characterization. These include the recently developed mass spectrometric ionization technique of fast atom bombardment as well as the more familiar chromatographic and electrophoretic procedures. Analyses of both the intact protein and its corresponding mixture of tryptic peptides oftentimes yield different but complementary information. Tryptic mapping techniques, which involve the separation of relatively small peptides can provide definitive data on the primary sequence. Knowledge concerning the tertiary structure of a protein is best obtained using techniques which rely upon the surface characteristics of the intact protein in addition to conventional spectroscopic techniques. These analytical procedures can be refined to optimize their particular resolving power toward the different properties of protein variants; be it electrophoretic, polar, or hydrophobic considerations. In this way, structural changes which are not observed under one set of conditions can be made apparent when analyzed using another protocol. It is thus possible to detect protein variants and chemically or enzymatically degraded species by a judicious combination of selected techniques.

By the time a protein pharmaceutical reaches the development stage and is considered for a clinical investigation a large amount of information about its physical and chemical properties has already been accumulated. The complexity of its structure has been established: its molecular weight, subunit composition, and glycosylation state. A number of suspected potential analogs and degraded species have been detected or postulated. In addition to analyzing a protein for characteristics dictated by its specific nature, routine analytical assays are performed; namely, amino acid analyses and amino- and carboxy-terminal sequencing. If necessary, disulfide assignments are made. A search is ordinarily conducted for oxidations and/or deamidations. In the case of recombinant proteins particular attention is paid to detecting signal sequences or proteolytically degraded species. In contrast, deletions and chemical modifications can be a consequence of proteins prepared by chemical synthesis. Lastly, tests are applied to determine if the protein is correctly folded into its native three dimensional structure. In addition, with recombinant proteins, sensitive immunological procedures are performed to ensure that host cell proteins have not been copurified with the protein of interest.

The following data has been assembled from the characterization of three very different proteins to illustrate the utility of a variety of tests which can be applied to ascertain the purity of protein pharmaceuticals: a small chemically synthesized protein consisting of two polypeptide chains held together by disulfide linkages (human relaxin), a recombinant protein of moderate size (human growth hormone) biosynthesized in bacterial cells, and a larger glycosylated recombinant protein (tissue plasminogen activator) secreted from mammalian cells.

Chemically Synthesized Human Relaxin

Recent elucidation of the primary structure of relaxin has revealed its homology to insulin particularly in its strikingly similar disulfide bond structure. The relaxin and insulin molecules are each composed of two nonidentical peptide chains linked by two disulfide bridges with an additional intrachain disulfide bridge in the smaller A-chain. The amino acid sequences of relaxin are known for a number of species including porcine (1), rat (2), sand tiger shark (3), spiny dogfish shark (4), human (5,6), skate (7), minke whale (8), and Bryde's whale (8). From protein sequencing data on the purified ovarian hormones and nucleotide sequence analysis data of cDNA clones (9,10) it appears that the relaxins are expressed as single chain peptide precursors with the overall structure: signal peptide/B-chain/C-peptide/A-chain. Since the sites of in vivo processing of human preprorelaxin are not yet identified, they were deduced by analogy to the processing of porcine and rat preprorelaxins (6). Hence, the human A-chain was chemically synthesized by solid phase methods as a 24 amino acid polypeptide and the B-chain as 33 amino acids in length.

Amino acid analysis can be used for the partial assessment of the primary structure of a polypeptide. With small peptides, such as relaxin, it is a useful technique for the determination of purity (11). At the very least, this method serves as a confirmation of the presence of the correct amino acids in the proper ratios. An analysis of the B-chain of human relaxin is shown in Table I. Since Thr and Ser are partially destroyed during acid hydrolysis, it is customary to perform extended hydrolyses and extrapolate to zero time. Conversely, the values for Val and Ile which are poorly hydrolyzed are best determined from the 72 h time hydrolysis. Inspection of the data for relaxin B-chain indicated that the value for Asx might be somewhat low. An amino terminal sequence analysis confirmed that the B-chain preparation did contain a secondary sequence of des Asp^{B1}-B-chain at a 5% level. The B-chain amino terminal sequence Asp^1-Ser^2 is particularly acid labile (12). Hence, the lack of an amino terminal Asp may be due to cleavage during the hydrofluoric acid treatment to remove the synthetic peptide from the solid phase rather than to an incomplete coupling during the chemical synthesis.

Reversed-Phase High Performance Liquid Chromatography (HPLC) of Relaxin. Since relaxin is composed of two polypeptide chains linked by disulfide bonds, it is possible to examine both the intact molecule and its composite chains after reduction(13). Relaxin samples can be analyzed by reversed-phase HPLC using a linear gradient of acetonitrile as described in the legend to Figure 1. This gradient resolves relaxin and its component A- and B-chains from each other and their minor variants. The profiles shown in Figure 1 are from a side fraction isolated from the combination reaction of the individual A- and B-chains to form intact relaxin. The profile obtained from analysis of the reduced material suggested that the impurities present in the side fraction were due to variants of the B-chain.

Tryptic Maps of Relaxin and Relaxin B-chain. Digestion of the A-chain of human relaxin with trypsin can theoretically result in the release of five fragments; that of the B-chain in the release of six fragments as illustrated in Table II. A typical tryptic map of relaxin B-chain is shown in Figure 2. The peptide was reduced and carboxymethylated with iodoacetic acid before enzymatic digestion. The peptide assignments were made after analysis of the peaks by amino acid hydrolysis for amino acid composition and confirmed by fast atom bombardment mass spectrometry (FAB-MS) as shown in Table III.

The presence of a des Asp^{B1} peptide is evident corroborating its previous discovery by amino acid and amino terminal sequence analyses of the intact B-chain. In addition, there is FAB-MS data for low levels of oxidation at the methionine residues present at positions B4 and B25.

Recombinant Human Tissue Plasminogen Activator (rt-PA)

In contrast to the relative simplicity of relaxin, rt-PA is a large glycosylated protein of approximately 65 kD. Perhaps its greater complexity is best illustrated by a comparison of the tryptic maps of these

Table I
Amino Acid Composition of Human Relaxin B-chain

Amino Acid	B-chain
CyA[a]	1.92 (2)[b]
Asx	0.93 (1)
Thr[c]	0.98 (1)
Ser[c]	3.90 (4)
Glx	3.97 (4)
Pro[a]	0 (0)
Gly	2.01 (2)
Ala	2.02 (2)
Val[d]	1.94 (2)
Met	1.82 (2)
Ile[d]	2.63 (3)
Leu	2.98 (3)
Tyr	0 (0)
Phe	0 (0)
His	0 (0)
Lys	1.93 (2)
Trp[e]	2.03 (2)
Arg	2.90 (3)

a   Performic acid oxidized sample results
b   Theoretical values are in parentheses
c   Extrapolation to zero time of hydrolysis
d   After a 72-h hydrolysis
e   Determined in the presence of thioglycolic acid

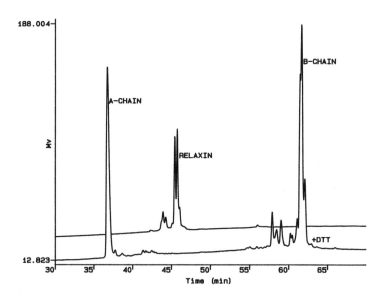

Figure 1. Reversed-phase HPLC chromatograms of human relaxin side fraction before and after reduction with dithiothreitol. The chromatography was performed on a Vydac C4 column using TFA-containing mobile phases, and eluted with an acetonitrile linear gradient from 18 to 50%.

Table II
Theoretical Tryptic Fragments

A-chain	
T_1 (A1-9)	pGlu-Leu-Tyr-Ser-Ala-Leu-Ala-Asn-Lys
T_2 (A10-17)	Cys-Cys-His-Val-Gly-Cys-Thr-Lys
T_3 (A18)	Arg
T_4 (A19-22)	Ser-Leu-Ala-Arg
T_5 (A23-24)	Phe-Cys

B-chain	
T_6 (B1-9)	Asp-Ser-Trp-Met-Glu-Glu-Val-Ile-Lys
T_7 (B10-13)	Leu-Cys-Gly-Arg
T_8 (B14-17)	Glu-Leu-Val-Arg
T_9 (B18-30)	Ala-Gln-Ile-Ala-Ile-Cys-Gly-Met-Ser-Thr-Trp-Ser-Lys
T_{10} (B31)	Arg
T_{11} (B32-33)	Ser-Leu

pGlu:Pyroglutamic acid

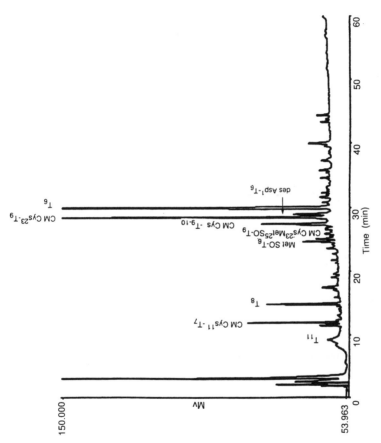

Figure 2. Tryptic map of human relaxin B-chain. The peptide was reduced with dithiothreitol and alkylated with iodoacetic acid before digestion with trypsin. The chromatography was performed on a Vydac C18 column using TFA-containing mobile phases, and eluted with an acetonitrile linear gradient.

Table III
Mass Spectral Analysis of Tryptic Map Peptides of
Human Relaxin B-chain

Peptide	Theoretical Mass	Observed
T_6	1136.4	1136.4
des Asp^1 – T_6	1021.4	1021.4
Met^4 SO – T_6	1152.4	1152.5
CM Cys^{11} – T_7	505.6	505.9
T_8	515.7	516.0
CM Cys^{23} – T_9	1453.9	1453.7
CM Cys^{23} Met^{25} SO – T_9	1469.9	1469.9
CM Cys^{23} – T_{9-10}	1609.9	1609.8
T_{11}	218.2	219.0

CM Cys, carboxymethyl cysteine
Met SO, methionine sulfoxide

proteins (Figure 3). While the map of relaxin could be generated from the intact molecule, retaining the disulfide linkages, it was necessary to first reduce and alkylate rt-PA in order to obtain digestion of the protein into its expected tryptic fragments.

Certain properties of rt-PA can lead to the generation of a simpler tryptic map. Human t-PA as produced in cell culture fluid (14) consists of one polypeptide chain. However, during isolation procedures (14,15) proteases arising from media components or lysed cells can lead to conversion into a two-chain variant consisting of a heavy chain (kringle region) and a light chain (protease region) which are connected by a disulfide bridge. The light chain contains the active site (16) and is homologous with other serine proteases. The heavy chain contains two kringle regions (17), a finger, and a growth factor domain (18). The two chains can be separated after reduction and carboxymethylation by gel filtration on a Sephadex G-75 superfine column (19). The tryptic maps of these isolated chains are depicted in Figure 4 and are more amenable to product characterization (19).

Recombinant Human Growth Hormone (rhGH)

Human growth hormone, a polypeptide of 191 amino acids, was first expressed in Escherichia coli using recombinant DNA techniques that resulted in production of the protein in the cytoplasm as a methionyl analog (20). Despite the reducing environment of the cytoplasm, the extracted and purified product contains the correct disulfide bonds and tertiary structure (21). In contrast, secretion methods allow the production in E. coli of hGH lacking the N-terminal additional methionine and with the correct disulfide bond formation (22). Whereas the secretion of correctly processed rhGH into the periplasm of E. coli represents a major process advantage in that the rhGH is easily extracted without the use of denaturants, a certain proportion of the molecules is cleaved to a two-chain form by an enzyme that may be located in the cell membrane. This two-chain form of rhGH was isolated by high resolution ion-exchange chromatography.

Reversed-phase HPLC of Two-chain rhGH and rt-PA.
For proteolysis to occur, the cleavage sites would be expected to be on the surface of the molecule and hence exposed to the column support. Therefore, it would be expected that two-chain species should be readily separable. Reversed-phase HPLC analysis in trifluoroacetic acid (TFA)-containing mobile phases (Figure 5) demonstrates that two-chain rhGH elutes immediately following the main peak (23). In contrast, a similar analysis of two-chain rt-PA results in a profile (Figure 6) in which the two-chain variant elutes before the one-chain species. However, when the growth hormone variant is chromatographed at neutral pH (Figure 7), it also elutes before its one-chain form (24). The apparent change in the order of elution can be related to the amount of denaturation of the two-chain form relative to the

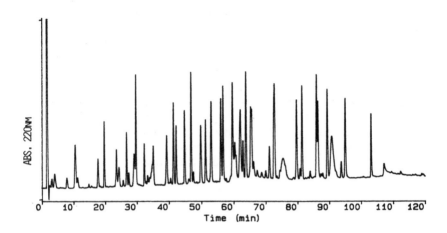

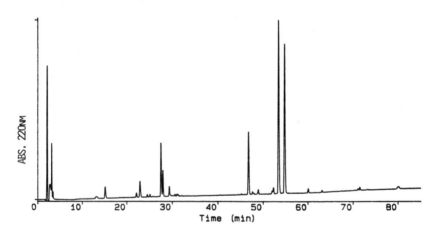

Figure 3. Comparison of rt-PA (reduced and carboxymethylated) with relaxin tryptic maps. The chromatography was performed as outlined in Figure 2 legend.

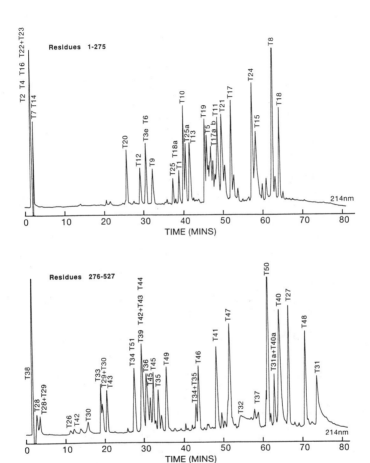

Figure 4. Tryptic map of reduced and carboxymethylated rt-PA heavy
and light chains. Chromatography was performed as outlined in
Figure 2 legend; upper panel: heavy chain; lower panel: light chain.
(Reproduced with permission from Ref. 19. Copyright 1989 Elsevier
Science Publishers.)

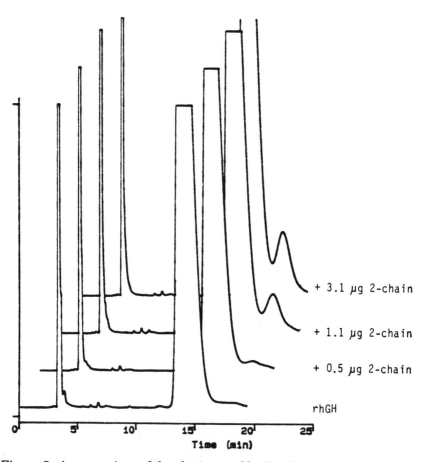

Figure 5. A comparison of the elution profiles for the analyses of rhGH with and without two-chain rhGH additions. The chromatography was performed using TFA-containing mobile phases and eluted with an acetonitrile linear gradient from 54 to 70% following a 10 min isocratic hold at 54% acetonitrile. (Reproduced with permission from Ref. 23. Copyright 1989 Munksgaard International Publishers.)

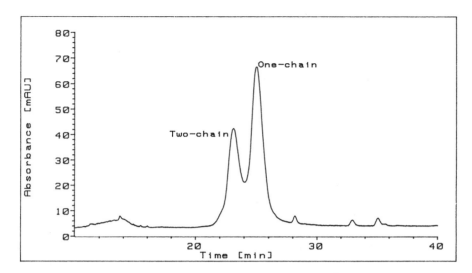

Figure 6. Reversed-phase HPLC of rt-PA. The chromatography was performed on a Vydac C$_4$ column using TFA-containing mobile phases and eluted with an acetonitrile gradient from 32 to 40%.

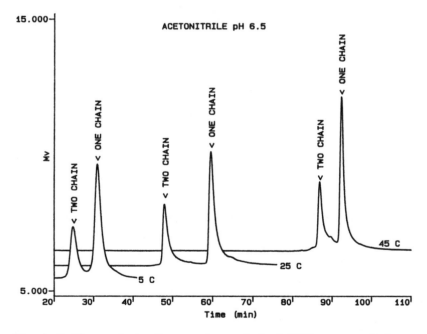

Figure 7. The analysis of one- and two-chain rhGH by reversed-phase HPLC at neutral pH. The chromatography was performed on a Polymer Laboratory PLRP-S column using phosphate-containing mobile phases, and eluted with an acetonitrile linear gradient.

native species during the chromatographic separation. In general it would be expected that a proteolytically clipped variant would elute before the one-chain form due to the greater polarity of the cleaved polypeptide chain, a result of the formation of charged end-groups. These conditions are apparently met for the chromatography of rhGH under less denaturing conditions (neutral pH) and for rt-PA (low pH, TFA) but not for rhGH under the low pH conditions. This difference can be related to the greater stability of rt-PA to denaturation, which can be seen from the resistance of rt-PA to trypsin digestion (25), a consequence of the presence of 17 disulfide bonds. By contrast, rhGH contains only two disulfides and is readily digested by trypsin in the absence of denaturants. Therefore, the later elution of the two-chain variant under more denaturing conditions (low pH, TFA) can be related to greater unfolding and the resultant exposure of more interior hydrophobic residues.

Separations as Influenced by pH. This concept of differential unfolding of proteins under varying mobile phase conditions was also illustrated in attempts to find conditions whereby rhGH could be separated from its methionine analog (Met-rhGH). When reversed-phase chromatograms were generated at a number of pH values different degrees of resolution were apparent (Figure 8). At low pH values little or no resolution was observed. As the pH was increased toward neutrality, clear resolution of the species was seen. Upon reaching basic conditions, resolution began to decrease. It could be rationalized that greater unfolding occurs at the extreme pH levels, obscuring any effect that the N-terminal methionine of the rhGH analog may have on binding of the hormone to the stationary phase. When the hormone is in its native state, the N-terminal methionine residue is exposed. This has been demonstrated to be the case with the porcine rGH analog by workers at Monsanto who have determined its crystal structure by x-ray diffraction techniques (26). It is thus possible for any increased binding contributed by the additional N-terminal amino acid present in Met-hGH to result in a separation of the two hGH species under conditions where the N-terminus is available for interaction with the column.
       That the N-terminal methionine is exposed to the solvent was also concluded from studies directed toward the reduction and alkylation of the disulfides of rhGH. The conditions that were found to be optimal for the derivatization of the four sulfhydryls of rhGH resulted in the production of a side reaction when the procedure was applied to the methionyl analog. This side product was identified as containing carboxymethyl-S-methionine at the amino terminal residue.

Separations as Influenced by Temperature. Variations in temperature were also investigated as a means of probing the selection of more refined reversed-phase conditions for the chromatography of rhGH. It is generally recognized that both the adsorption of proteins onto the nonpolar solid phase used in reversed-phase chromatography and the harsh conditions required for elution can lead to denaturation of proteins (27). Benedek et al. have shown that acetonitrile is a more denaturing organic modifier than propanol in protein chromatography by reversed-phase HPLC (27), so that the degree of resolution can be further investigated by use of different solvents. Therefore, both denaturing, low pH and acetonitrile organic modifier, and less denaturing, neutral pH and n-propanol organic modifier, conditions were studied (28). The ability of low pH to unfold the protein is demonstrated by the temperature plots presented in Figure 9. Although isocratic conditions are necessary for the investigation of kinetic parameters, the use of a shallow gradient can be utilized to scan the effect of a wide range of mobile phase conditions on a given separation. In fact, due to the pronounced effect of temperature and mobile phase conditions on the retention times of proteins, the use of isocratic conditions can allow the examination of only a narrow range of conditions.

In reversed-phase HPLC of organic molecules and low molecular weight peptides the effect of an increase in temperature is a corresponding decrease in retention time. However, many proteins are known to unfold with an increase in temperature resulting in exposure of hydrophobic groups. Thus, in a chromatographic separation of a protein, if an increase in temperature does lead to a significantly greater degree of unfolding, longer retention times may result. This phenomenon was observed both at low (Figure 9) and high pH in the chromatography of rhGH (28) when acetonitrile was used as an organic modifier. However, the rate of increase plateaued at higher temperatures so that above 45° to 60°C a decrease in retention time was observed. This observation is consistent with a reduction in the rate of denaturation of the protein at higher temperatures where the protein may be maximally unfolded. While it can be seen in Figure 9 that acetonitrile, a denaturing solvent, led to greater retention times with increased temperatures at low pH, the effect was even more pronounced at neutral pH (28), where the molecule is presumably in a more native state at the initial temperature point. The use of n-propanol gave a dramatic difference in the temperature effects (28). At both pH values the plot was similar to the high temperature values for the acetonitrile study. Thus, the less denaturing organic solvent resulted in the normal observation of reduced retention times at higher temperatures. A similar study done with rt-PA led to the conclusion that this molecule was fairly rigid since with the low pH mobile phase containing acetonitrile it did not tend to be more retained on the column at higher temperatures.

An attempt to find the best conditions for separating the one- and two-chain forms of rhGH was made using temperature as a probe with mobile phases of varying denaturing capacities. It was found that the optimal conditions for resolution were neutral pH, acetonitrile, and 25°C. As an example, Figure 7 shows the result for the use of temperature to optimize the separation at pH 6.5. Further evidence for the unfolding power of low pH and the acetonitrile organic modifier is uncovered when

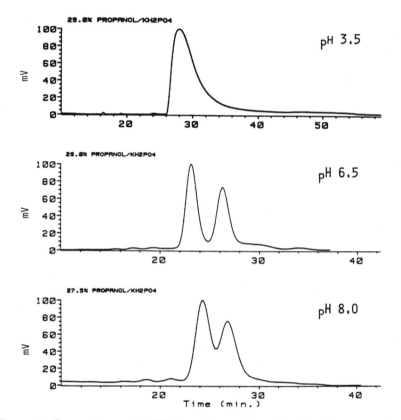

Figure 8. Separations of rhGH from its methionyl analog by reversed-
phase HPLC. The chromatography was done at the indicated pHs
using phosphate-containing mobile phases with propanol organic
modifier on a Vydac C_4 column. Elutions were run in an isocratic
mode.

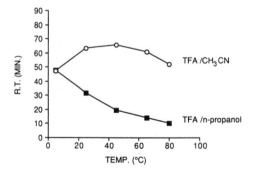

Figure 9. Effect of temperature and organic modifier at low pH on the
reversed-phase chromatography of rhGH. Elutions were done with a
shallow linear gradient of 0.2%/min.

the data obtained at 5°C is examined as follows. At 5°C any denaturation due to temperature is minimized. If the starting acetonitrile concentration is low (28%) the molecules are less unfolded than if the starting concentration is high (38%). As the acetonitrile concentration is increased to elute the proteins from the column more unfolding occurs in the molecule which was loaded at a lower acetonitrile concentration, binding is increased, resident time on the column is longer (110 min at 28% starting acetonitrile concentration as opposed to 50 min at 38%), and elution requires a greater acetonitrile concentration (57% at 28% starting acetonitrile concentration as opposed to 50% at 38%). No matter what the starting conditions are, the two-chain form requires greater acetonitrile concentrations for elution at low pH, most likely due to its greater flexibility and hence greater unfolding potential.

It was expedient to develop a sensitive assay for the detection of two-chain rhGH since this variant is a result of biosynthesis in E. coli and not found in pituitary-derived material. In contrast, two-chain rt-PA is a consequence of the activation process required for the dissolution of a blood clot. Its quantitation is important for a different reason: to demonstrate consistency in the manufacturing process.

Detection of Deamidation

The strategy used for the detection of deamidation of these three proteins had to be altered due to their particular characteristics. Neither isoelectric focusing (IEF) nor tryptic mapping, which were effective in monitoring the deamidation of rhGH (29) produced any evidence for the deamidation of relaxin. Since deamidation of asparagine side chains is commonly seen in proteins (30), position A8 was of interest. The T_1 peptide was chemically synthesized with an aspartic acid residue at that position so its elution time in the relaxin tryptic map could be determined. It eluted in a position completely devoid of any absorbing material in the tryptic map of a typical relaxin sample (13) which has a detection limit of approximately 2%, indicating that this asparagine residue is not particularly susceptible to deamidation. This is not surprising since it is followed by lysine, an amino acid that does not favor the formation of the cyclic imide which is an intermediate in the deamidation reaction (31). The presence of an aspartic acid residue adjacent to lysine probably would not affect the rate of trypsin digestion at that position since the sequence $Asp^{169}Met^{170}Asp^{171}Lys^{172}$ in hGH is completely released under similar conditions (32).

A different problem exists in the case of rt-PA. Its IEF pattern is shown in Figure 10. The great number of bands seen is due to the heterogeneity of the carbohydrate moieties. Any deamidation is hence hidden in this complex pattern. A great deal of effort is necessary to identify any deamidations in this protein. It was necessary to incubate rt-PA at pH 5.5 for ten months at 30°C to achieve a measurable level of degradation. The molecule was then reduced and alkylated before separating the various species by reversed-phase chromatography. A peak not apparent in the reversed-phase HPLC profile of reference material was isolated and digested with trypsin. By this procedure it was possible to identify deamidated peptides (they usually elute slightly later than the native peptide giving rise to doublet or triplet peak patterns). Two such peptide profiles were observed, T_5 and T_8 (Figure 11). The FAB-MS analysis on the individually isolated T_5 doublet peaks confirmed the deamidation, T_5:1332 amu; deamidated T_5:1333 amu. Due to the complexity of the tryptic map (with at least 60 peptides), it will be very difficult to identify all sites of deamidation in this manner and the comparison of a given map relative to that of a reference material will not allow the detection of low levels of a variant (less than 10%).

A procedure that can determine if deamidation has occurred in the rt-PA molecule makes use of bovine protein carboxyl methyltransferase (PCMT), an enzyme that methylates the free carboxyl group present at atypical isoaspartyl linkages (33), a product that is indicative of deamidation (34). The ratio of isoaspartate to aspartate formed due to deamidation is generally about 3:1 (33,35). The methylation of the isoaspartyl peptide or protein by PCMT has been reported to be stoichiometric (36). Methylation by PCMT is measured by incubating the deamidated peptide or protein with the radiolabelled methyl donor, S-adenosyl-L-[3H-methyl]-methionine and subsequently monitoring the incorporated radiolabel in a liquid scintillation counter. This method has been used successfully in corroborating the evidence for deamidation of hGH (29) occurring during in vitro aging (35). However, if the deamidated residues are buried in the interior of the protein, it is likely that the enzyme will not be able to methylate such residues. When native rt-PA is exposed to this enzyme, very little methylation is evident (less than 0.05 residues; Figure 12). This result could, however, be due to steric hindrance of the enzyme due to the highly cross-linked structure of rt-PA. If the rt-PA is first reduced and carboxymethylated to break the disulfides and destabilize the structure, an increase in methylation by PCMT is seen. Digesting the reduced and alkylated molecule with trypsin leads to an even greater methylation of rt-PA consistent with the hypothesis that disruption of the 3D-structure does indeed allow access of the enzyme to the isoaspartyl residues. Treatment of rt-PA with alkali (pH 8.0) which is proposed to result in the deamidation of proteins resulted in methylation of almost 0.6 residues (Figure 12). The increase of 0.22 residues of methyl-accepting capacity in the deamidated sample relative to the control (Figure 12: trypsin digest of reduced and carboxymethylated rt-PA) can be attributed to deamidation. This value indicates that rt-PA is a relatively stable protein as compared to Met-rhGH with 0.35 residues of methyl-accepting capacity when treated similarly (35) especially since there are a greater number of asparagine residues present in rt-PA. An explanation for this stability could reside in the highly cross-linked structure of rt-PA as well as in the presence of the bulky carbohydrate side chains. Kossiakoff (37)

STD   rt-PA

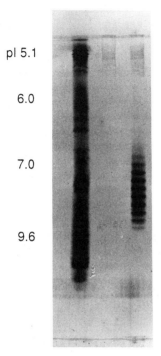

pl 5.1

6.0

7.0

9.6

Figure 10. Isoelectric focusing of native rt-PA.

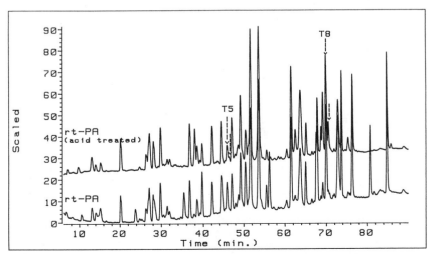

Figure 11. Comparison of the tryptic maps of reduced and
carboxymethylated rt-PA standard and acid-treated preparations.
Chromatography was performed as outlined in Figure 2 legend.

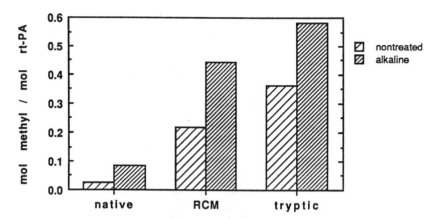

Figure 12. Comparison of methyl-accepting capacity of nontreated and alkali-treated rt-PA. To label rt-PA, 1 nmol of rt-PA was combined with 125 pmol of PCMT along with 10 µl of 500 dpm/pmol of 1 µM tritiated S-adenosylmethionine. The mixture was incubated for 40 minutes at 30°C.

Table IV

Comparison of the Major Characteristics of

Relaxin, rhGH, and rt-PA

Molecular Property	Relaxin	rhGH	rt-PA
Disulfide assignment	Difficult	Easy	Impossible
Deamidation	None	As expected	Difficult to assay
Proteolysis to two-chain	Does not apply	Due to host cell	Integral property
3-D structure	Leads to high[a] yield during chain combination	Flexible	Rigid
Glycosylation	No	No	Yes

[a] Higher than expected on a statistical basis.

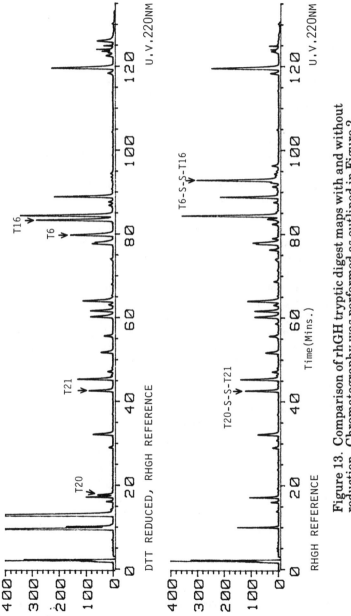

Figure 13. Comparison of rhGH tryptic digest maps with and without reduction. Chromatography was performed as outlined in Figure 2 legend.

showed that for deamidation to occur a distinct local conformation and
hydrogen-bonded structure of the amide group was required.

Disulfide Assignments

Since both relaxin and rhGH can be completely digested by trypsin without
first reducing their disulfide linkages, it is possible to identify the peptides
which are involved in each linkage. The tryptic map of rhGH is shown in
Figure 13 with the cystine-containing peptides indicated in the usual
manner; the theoretical tryptic fragments are numbered beginning at the
amino terminal residue. The map of the tryptic digest treated with
dithiothreitol is also reproduced in Figure 13. Of particular interest is the
similar elution times of T_{21} in this map with that of T_{20}-T_{21} in the map of
the untreated digest. Hence, identification of peptides by chromatographic
retention times may be a dangerous practice.

The assignments of the cystine-containing peptides of relaxin were
made in a similar manner, with one caveat. When the enzymatic digestion
was conducted at pH 8, all possible cystine-containing peptide pairs were
observed. This observation indicated that a disulfide exchange was
occurring in the basic medium. Adjusting the buffer pH to 7.2 prevented
this exchange reaction and permitted the proper assignments, T_2-T_7 and
T_5-T_9, to be made (13).

As previously noted, rt-PA is resistant to tryptic digestion if the
disulfide linkages are left intact. Hence, to date, no complete direct
assignment of these linkages have been made. Another difference which
exists in the rt-PA molecule is the presence of an odd number of cysteine
residues resulting in at least one free sulfhydryl group in the protein.

Conclusions

The above discussions have shown how selected analytical techniques can
be applied to vastly different proteins to solve a myriad of problems. These
include routine assays: amino acid and sequencing analyses; specialized
techniques: FAB-MS and IEF; conventional techniques refined to improve
their utility: reversed-phase HPLC using different pHs, organic modifiers,
and temperatures; and chemical and enzymatic modifications. The latter
two procedures have been shown to be effective not only in elucidating
primary structure but also in probing the conformation of proteins.
Table IV is a summary of the information that has been compiled for
human relaxin, growth hormone, and rt-PA.

Acknowledgments
The authors wish to thank the following people for their work in the
development of the analytical procedures utilized in this study: Ida
Baldonado for the reversed-phase HPLC and tryptic map of relaxin;
Dr. John Stults for FAB-MS analyses; John Battersby for the reversed-

phase HPLC of growth hormone; Victor Ling for the reduction and carboxymethylation of growth hormone; and Reed Harris for amino acid analyses. One of us, AWG, would like to especially thank Dr. Dana Aswad for a brief internship in his laboratory and for his gift of purified bovine protein carboxyl methyltransferase.

Literature Cited

1. James, R.; Niall, H.; Kwok, S.; Bryant-Greenwood, G. Nature 1977, 267, 544-546.
2. John, M.J.; Borjesson, B.W.; Walsh, J.R.; Niall, H.D. In Relaxin; Bryant-Greenwood, G.D.; Niall, H.D.; Bryant-Greenwood, F.C., Eds.; Elsevier/North-Holland: New York, 1981; pp 37-43.
3. Gowan, L.K.; Reinig, J.W.; Schwabe, C.; Bedarkar, S.; Blundell, T.L. FEBS Lett. 1981, 129, 80-82.
4. Büllesbach, E.E.; Gowan, L.K.; Schwabe, C.; Steinetz, B.G.; O'Byrne, E.; Callard, I.P. Eur. J. Biochem. 1986, 161, 335-341.
5. Hudson, P.; Haley, J.: John, M.; Cronk, M.; Crawford, R., Haralambidis, J.; Tregear, G.; Shine, J.; Niall, H. Nature 1983, 301, 628-631.
6. Hudson, P.; John, M.; Crawford, R.; Haralambidis, J.; Scanlon, D.; Gorman, J., Tregear, G.; Shine, J.; Niall, H. EMBO 1984, 3, 2333-2339.
7. Büllesbach, E.E.; Schwabe, C.; Callard, I.P. Biochem. Biophys. Res. Commun. 1987, 143, 273-280.
8. Schwabe, C.; Büllesbach, E.E.; Heyn, H.; Yoshioka. J. Biol. Chem. 1989, 264, 940-943.
9. Haley, J.; Hudson, P.; Scanbon, D.; John, M.; Cronk, M.; Shine, J.; Tregear, G.; Niall, H. DNA 1982, 1, 155-162.
10. Hudson, P.; Haley, J.; Cronk, M.; Shine, J.; Niall, H. Nature 1981, 291, 127-131.
11. Chance, R.E.; Kroeff, E.P.; Hoffman, J.A. In Insulins, Growth Hormone and Recombinant DNA Technology; Gueriguian, J.L., Ed.; Raven Press: New York, 1981, pp 71-86.
12. Tsung, C.M.; Fraenkel-Conrat, H. Biochemistry 1965, 4, 793-800.
13. Canova-Davis, E.; Baldonado, I.P.; Teshima, G.M. J. Chromatog. submitted.
14. Rijken, D.C.; Collen, D. J. Biol. Chem. 1981, 256, 7035-7041.
15. Wallen, P.; Ranby, M.; Bergsdorf, N.; Kok, P. Prog. Fibrinolysis 1981, 5, 16-23.
16. Rijken, D.C.; Wijngaards, G.; Zaal-de Jong, M.; Welbergen, J. Biochim. Biophys. Acta 1979, 580, 140-153.
17. Pennica, D.; Holmes, W.E.; Kohr, W.J.; Harkins, R.N.; Vehar, G.A.; Ward, C.A.; Bennett, W.F.; Yelverton, E.; Seeburg, P.H.; Heyneker, H.L.; Goeddel, D.V.; Collen, D. Nature 1983, 301, 214-221.
18. Banyai, L.; Varadi, A.; Patthy, L. FEBS Lett. 1983, 163, 37-41.
19. Chloupek, R.C.; Harris, R.J.; Leonard, C.K.; Keck, R.G.; Keyt, B.A.; Spellman, M.W.; Jones, A.J.S.; Hancock, W.S. J. Chromatogr. 1989, 463, 375-396.

20. Goeddel, D.V.; Heyneker, H.L.; Hozumi, T.; Arentzen, R.; Itakura, K.; Yansura, D.G.; Ross, M.J.; Miozzari, G.; Crea, R.; Seeburg, P.H. Nature 1979, 281, 544-548.
21. Jones, A.J.S.; O'Connor, J.V. In Hormone Drugs, Proceedings of the FDA-USP Workshop on Drug Reference Standards for Insulins, Somatotropins and Thydroid-Axis Hormones; Gueriguian, J.L., Ed.; U.S. Pharmacopeial Convention : Bethesda, MD, 1982; pp 335-351.
22. Gray, G.L.; Baldridge, J.S.; McKeown, K.S.; Heyneker, H.L.; Chang, C.N. Gene 1985, 39, 247-254.
23. Canova-Davis, E.; Baldonado, I.P.; Moore, J.A.; Rudman, C.G.; Bennett, W.F.; Hancock, W.S. Int. J. Peptide Protein Res. In press.
24. Canova-Davis, E.; Baldonado, I.P.; Basa, L.J.; Chloupek, R.; Doherty, T.; Harris, R.J.; Keck, R.G.; Spellman, M.W.; Bennett, W.F.; Hancock, W.S. In Peptides, Chemistry and Biology; Marshall, G.R., Ed.; Proceedings of the Tenth American Peptide Symposium; ESCOM : Leiden, 1988; pp 376-378.
25. Keyt, B.A.; Teshima, G.; Harris, R.J.; Jones, A. Tenth International Symposium on Column Liquid Chromatography; San Francisco, CA., 1986, p 813.
26. Abdel-Meguid, S.S.; Shieh, H.-S.; Smith, W.W.; Dayringer, H.E.; Violand, B.N.; Bentle, L.A. Proc. Natl. Acad. Sci. USA 1987, 84, 6434-6437.
27. Benedek, K.: Dong, S.; Karger, B.L. J. Chromatog. 1984, 317, 227-243.
28. Teshima, G.; Wu, S.-L.; Hancock, W.S. Manuscript in preparation.
29. Hancock, W.S.; Canova-Davis, E.; Chloupek, R.C.; Wu, S.-L.; Baldonado, I.P.; Battersby, J.E.; Spellman, M.W.; Basa, L.J., Chakel, J.A. In Therapeutic Peptides and Proteins : Assessing the New Technologies; Banbury Report 29; Cold Spring Harbor Laboratory, 1988; pp 95-109.
30. Robinson, A.B.; Rudd, C.J. In current Topics in Cellular Regulation; Horecker, B.L.; Stadtman, E.R., Eds.; Academic Press : New York, 1974; pp 247-295.
31. Bodanszky, M.; Kwei, J.Z. Int. J. Peptide Protein Res. 1978, 12, 69-74.
32. Canova-Davis, E; Chloupek, R.C.; Baldonado, I.P.; Battersby, J.E.; Spellman, M.W.; Basa, L.J.; O'Connor, B.; Pearlman, R.; Quan, C.; Chakel, J.A.; Stults, J.T.; Hancock, W.S. American Biotechnology Laboratory 1988, 6, 8-17.
33. Murray, E.D., Jr.; Clarke, S. J. Biol. Chem. 1984, 259, 10722-10732.
34. Bornstein, P.; Balian, G. Meth. Enzymol. 1977, 47, 132-145.
35. Johnson, B.A.; Shirokawa, J.M.; Hancock, W.S.; Spellman, M.W.; Basa, L.J.; Aswad, D.W. J. Biol. Chem. 1989, 264, 14262-14271.
36. Aswad, D.W. J. Biol. Chem. 1984, 259, 10714-10721.
37. Kossiakoff, A.A. Science 1988, 240, 191-194.

RECEIVED December 20, 1989

Chapter 6

Analytical Chemistry of Therapeutic Proteins

R. M. Riggin and N. A. Farid

Lilly Research Laboratories, Eli Lilly and Company, Indianapolis, IN 46285

The manufacturing process for biomolecules intended for therapeutic use must include extensive quality control measures to insure the production of a safe and effective product which meets rigorous standards of quality and batch-to-batch consistency. While this requirement applies to both small molecules and proteins, the manufacturing and structural complexity of proteins dictates the use of a much more extensive battery of analytical tests, many of which are quite complex and not routinely conducted in conventional analytical laboratories.

As with small molecules, the key analytical parameters to be addressed for therapeutic proteins on a lot-to-lot basis are identity, purity, and potency. In addition, a rigorous proof of structure is required for the reference standard lot to be used for routine analysis of production batches.

Characterization of Protein Structure

The complexity of quality control for proteins, as compared to small molecules, is most evident in the requirements for proof of structure. Many small molecules can be fully characterized using a few spectroscopic techniques (e.g., NMR, IR, mass spectrometry, and UV) in conjunction with an elemental analysis. However, proving the proper structure for a protein is much more complex because 1) the aforementioned spectroscopic techniques do not provide definitive structural data for proteins, and 2) protein structure includes not only molecular composition (primary structure) but additionally, secondary, tertiary, and, in some cases, quaternary features. Clearly, no single analytical test will address all of these structural aspects; hence a large battery of tests is required.

0097–6156/90/0434–0113$06.00/0
© 1990 American Chemical Society

The analytical techniques frequently employed to determine the primary structure of a protein are listed in Table I. In general, most or all of these tests will be used for proof of structure, with a subset or streamlined versions of these techniques used to confirm identity on a lot-to-lot basis.

Table I. Techniques Used for Determination of Primary Structure

---

Gene (DNA) sequencing
Edman degradation/sequence analysis
C-terminal sequencing
Peptide mapping (characterization of disulfide linkages)
Raman Spectroscopy (disulfide linkages)
FAB/MS; PD/MS; MS/MS
Amino acid composition
Carbohydrate composition/structure

Sequencing of the DNA comprising the structural gene is routinely conducted and also serves to confirm primary structure of the protein; however, any post-translational modifications to the protein will not be evident from the gene sequence. The linear sequence of amino acids within a protein is most frequently determined by Edman degradation of the protein as well as peptide fragments generated from the protein using proteolytic enzymes. Recently, fast atom bombardment/mass spectrometry (FAB/MS), plasma desorption/mass spectrometry (PD/MS), and tandem mass spectrometry (MS/MS) have become widely used for the characterization of proteins and peptides. These techniques offer the potential for more sensitive and rapid structural characterization, and can be used to determine post-translational modifications. Disulfide bond formation is most frequently determined by peptide mapping techniques. Raman spectroscopy can also be used for ascertaining disulfide bond arrangements, and is most useful when an authentic standard is available for spectral comparision. Amino acid composition is generally not of great value in determining primary structure, but can be used to confirm the presence of unusual amino acids (e.g., gamma-carboxy glutamic acid). Complete characterization of the carbohydrate portion of glycoproteins is a very formidable task due to the heterogeneity of glycoproteins (i.e., since glycosylation is not under direct genetic control, a variety of carbohydrate structures generally exist for any particular glycoprotein product). Recent advances in analytical technology have made more detailed characterization possible, but complete routine characterization of carbohydrates is still a formidable challenge.

Analytical techniques used to determine secondary and tertiary structure are summarized in Table II. Generally, these techniques provide complementary information with evidence of structure arising

from a systematic evaluation of data obtained over a wide range of
solvent conditions (e.g., denaturant concentrations) using a combination of several techniques. X-ray crystallographic data are frequently not possible to obtain, due to difficulties in identifying
conditions for protein crystallization. Furthermore, in many cases
the conditions eventually identified for crystallization may be very
different than conditions used in the production process, hence
making the X-ray crystallographic data of limited value for determination of solution-state secondary and tertiary structure. In
general, the most direct evidence for proper tertiary structure is
the specific biological activity of the protein, since proper biological functioning of the molecule is dependent on the three-
dimensional structure of the molecule.

Table II. Techniques Used for Determination of Secondary and
           Tertiary Structure

SECONDARY STRUCTURE
    - Circular dichroism (CD)
    - Optical Rotatory Dispersion (ORD)
    - Fluorescence (time resolved/phase modulated)
    - Raman
    - Infrared (IR)

TERTIARY STRUCTURE
    - Two dimensional NMR
    - X-ray crystallography
    - Specific biological activity (in-vivo/in-vitro)

Routine Confirmation of Product Identity

Once the structural features of a reference standard of the desired
protein have been well characterized, lot-to-lot confirmation of
identity can be conducted using a carefully selected group of tests,
wherein the lot undergoing analysis is compared to the reference
standard. Tests commonly employed for this purpose are listed in
Table III. Peptide mapping is perhaps the most powerful and univer-
sally used technique since it provides relatively specific confirmation of correct primary sequence and, when non-reducing conditions
are employed, can be used to confirm correct disulfide bond formation. Tertiary structure is difficult to address directly on a
routine (lot-to-lot) basis, and the presence of correct biological
activity is often used as evidence that the correct tertiary
structure is maintained.

Due to the structural complexity of proteins, proper confirmation of product identity generally requires the use of several
different techniques in parallel. For example, use of RP-HPLC and
capillary electrophoresis in parallel provides a powerful means for
proving identity of a particular peptide fragment since these two
techniques exhibit independent elution profiles for typical peptide
digests (1); with RP-HPLC separating on the basis of hydrophobicity and capillary electrophoresis separating on the basis of
charge.

Using the arsenal of analytical methods currently available,
one can reliably confirm protein identity, even when seemingly minor
structural differences exist between two proteins. For example,
native sequence human growth hormone (hGH) and N-methionyl hGH, or
met-hGH (both of which are commercial products), can be readily
distinguished by RP-HPLC retention (2) and tryptic peptide mapping
(3). In addition, as illustrated in Figure 1, these two proteins
can be distinguished on the basis of their differing chemical
properties. As shown in Figure 1, met-hGH is oxidized by dilute
hydrogen peroxide at a much greater rate (approximately 4-5 fold
faster) than natural sequence hGH. This difference is believed to
be due to the rapid oxidation of the N-terminal methionine to a
sulfoxide, since we have demonstrated that oxidation of methionine
groups in hGH is a significant degradation pathway in commercial
products (4). Control solutions of hGH and met-hGH not containing
hydrogen peroxide exhibited no degradation over the 9-hour time
period of the experiment, thereby confirming that oxidation was the
main degradative pathway in this study.

Human, porcine, and bovine insulins can also be readily distinguished from one another using either peptide mapping or RP-HPLC
separation of the intact proteins (5).

Table III. Techniques Used for Routine Confirmation of Protein
           Product Identity

Chromatographic retention (e.g., RP-HPLC)
Electrophoretic mobility (size/charge based separations)
Peptide map (reduced/non-reduced)
Amino acid composition
Carbohydrate composition
Specific biological activity

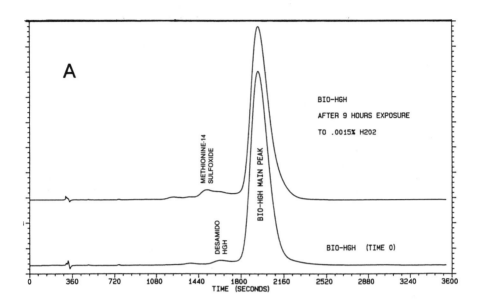

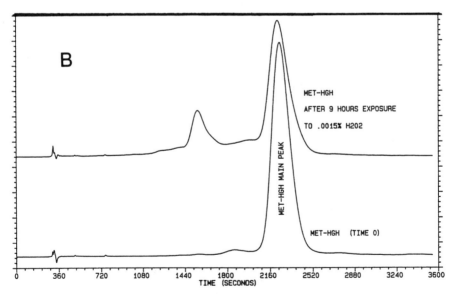

Figure 1.   RP-HPLC Profiles for met-hGH (Panel A) and hGH (Panel B)
Before and After Exposure to Aqueous Hydrogen Peroxide.
See Reference 2 for chromatographic conditions.

STRATEGIES/TECHNIQUES FOR ASSESSING PURITY OF PROTEINS

While the concept of purity is rather straightforward for small molecules, definition of purity for a protein product can be difficult. The difficulty in defining purity arises for several reasons: 1) some products, glycoproteins in particular, exist in a multitude of biologically active forms, many or most of which are quite acceptable therapeutically, 2) not all chemically equivalent forms of a protein may be acceptable therapeutically due to differences in aggregation state, etc., and 3) the level of purity observed for a given product is extremely dependent on the specific analytical technique used. Whenever one is faced with the task of interpreting purity data for a given product, the impact of these issues must be recognized and taken into consideration. For example, a pituitary hGH product which was stated to be 95% pure (determined by immunooassay for prolactin and similar contaminants) was determined to be only about 65% pure when assayed by reversed-phase HPLC (2).

Purity of therapeutic protein products is perhaps best addressed by dividing impurities into two categories: 1) contaminants structurally unrelated to the protein of interest and 2) components chemically related to the product of interest. A representative list of potential "unrelated contaminants" is shown in Table IV.

Table IV. Potential Contaminants Not Structurally Related to the Protein of Interest

---

Host cell proteins
Residual DNA
Media components (growth factors, protease inhibitors, trace elements, etc.)
Recovery/purification reagents (salts, detergents, urea enzymes, filter aids, etc.)
Endotoxin
Live microbes
Live viruses
Mycoplasma

Assays for media components and purification reagents are generally specific to the particular item used and will not be discussed further here. Assays for endotoxin, live virus, mycoplasma, and live microbes are relatively standard (USP, CFR, or other regulatory procedures) and likewise will not be discussed in detail. The determination of trace levels of host cell proteins in therapeutic proteins can be a formidable task, because most commonly used procedures for detection of proteins will encounter interference from the protein product itself. Furthermore, the wide range of potential host cell protein contaminants makes determination of each individual contaminant impossible. The approach usually employed for detection of host cell proteins has been described (6). In this process, host cell proteins are obtained from a cell colony identical to the production strain, but lacking the gene for the desired protein product. This host cell protein mixture is then used to develop antisera (in an appropriate animal) which will recognize the host cell proteins. The development of suitable antisera is usually a multiple step process, since care must be taken to insure that a wide range of host cell proteins is recognized, not simply one highly antigenic protein. Once suitable antiserum is available, this is used in an ELISA or other immunoassay to quantify levels of host cell proteins in the product (using the mixture of host cell proteins as the assay standard).

The detection of trace levels of residual DNA in protein products is a formidable challenge, primarily due to the extremely low detection limits required (e.g., 10-100 pg/dose). Probe hybridization, using a radiolabeled DNA probe derived from host cell DNA, has been in the method most commonly used, due to the extraordinary sensitivity of the assay. However, it is important to remember that this assay will recognize only DNA complementary to the labeled probe (i.e., other forms of DNA will not be recognized in the assay). Several techniques are now available for general detection of DNA at relatively low levels, and these methods are expected to come into more widespread use in the future. One such method is based on the use of DNA binding proteins in a format similar to an ELISA assay.

Typical components structurally related to the protein product are listed in Table V. The particular components of interest for any given protein will depend on the specific chemical and physical properties (e.g., presence or absence of methionine groups, etc.)

Table V.   Typical Components Structurally Related to the
           Protein of Interest

Deamidated forms
Methionine sulfoxides
Proteolytically clipped forms
N-Terminal variants
Disulfide isomers
Dimers/aggregates (covalent/non-covalent)
Glycosylation variants

Techniques commonly used to quantify such components are listed
in Table VI. Generally, both chromatographic and conventional electrophoretic techniques are applied to a given protein product to
insure complete characterization. However, chromatographic techniques are preferable in most cases due to ease of automation and
quantitation. Rapid progress in the commercial development of
capillary electrophoresis makes this technique a likely candidate
for routine use in the near future. Examples demonstrationg the
use of reversed-phase HPLC, anion exchange HPLC, and capillary
electrophoresis for quantifying deamidated forms of hGH and insulin
have been published (2,4,5).

Table VI. Techniques Used to Quantify Components Structurally
           Related to the Protein of Interest

TECHNIQUE	TYPE OF COMPONENTS DETERMINED
CHROMATOGRAPHY	
Ion exchange/chromatofocusing	Deamidated forms, glycosylation variants
Reversed-phase/hydrophobic interaction	N-terminal variants, glyco-sylation variants, disulfide isomers, proteolytic clips
Size exclusion	Dimers/aggregates
ELECTROPHORESIS	
SDS-PAGE	Proteolytic clips, dimers/ aggregates (covalent)
IEF	Deamidated forms, glycosylation variants
CAPILLARY ZONE ELECTROPHORESIS	Deamidated forms, N-terminal variants, proteolytic clips

The introduction of "fast HPLC" has proven to be particularly
valuable in protein analysis. As stated earlier, assay time in
RP-HPLC analysis of proteins is typically long compared to that for
smaller organic molecules. We have evaluated the use of 0.6-cm ID
x 4-cm columns packed with 3-um particles in the analysis of insulin
by RP-HPLC for potency determination, related substances, and in
peptide mapping (7). The use of the "fast column" allows considerable savings (40-60%) in analysis time, compared to the regular
(0.46-cm ID x 25-cm) columns, without loss in resolving power.
Also, the use of the "fast columns" did not require modifications

in eluent compositions beyond those that are typically part of the routine adjustments carried out by the analyst to meet system suitability parameters. As expected, the use of short columns packed with 3-um particles results in reduced solvent consumption and reduced mobile phase re-equilibration time in gradient elution. Data obtained in the HPLC analysis of insulin formulations, utilizing a 25-cm column and a 4-cm (3-um particle) column, are summarized in Table VII. The data clearly show the equivalent performance of both columns.

Table VII. Comparison of a 25-cm Column and a 4-cm (3-um) Column
            in the Analysis of Insulin Formulations(a)

Insulin Formulation	HPLC Potency, U/mL(b) Mean ± S.D. (n=6)		% Related Substances(c) Mean ± S.D. (n=6)	
	4 cm	25 cm	4 cm	25 cm
Regular	101.0 (0.4)	100.1 (1.1)	1.6 (0.3)	1.6 (0.2)
NPH	100.5 (0.7)	100.5 (0.9)	1.5 (0.5)	1.6 (0.6)
Zinc suspension	100.2 (0.9)	100.1 (0.9)	0.8 (0.2)	0.8 (0.2)

(a) For chromatographic analysis conditions, see ref. (5).
(b) Isocratic RP-HPLC method.
(c) Gradient RP-HPLC method.

    Peptide mapping has been widely used for characterization of protein products, and is almost always used for confirmation of identity. Generally, peptide mapping is not a very sensitive technique for detection of impurities; in our experience, derivatives present at levels below a few percent are not normally detected in peptide maps, whereas HPLC of the intact protein can generally detect 0.1% of such derivatives. However, in relatively simple maps such as the V-8 protease digest of insulin, relatively good limits of detection can be achieved (5).

    Correct chemical composition of a protein product does not always describe overall purity, since the existence of physically aggregated forms must also be considered. For example, a dimeric (non-covalent) form of hGH has been isolated, using high performance size exclusion chromatography (HPSEC) under non-denaturing conditions, which exhibits only about 20% of the biological activity of monomeric hGH (8). This dimer was found to be non-covalent as evidenced by the SDS-PAGE and RP-HPLC profiles; both of these techniques tend to dissociate noncovalent aggregates. The tryptic peptide map of this component was identical to that of monomeric hGH. Clearly, the use of HPSEC under non-dissociating conditions is an important measure of purity for this particular product.

Techniques/Strategies for Determining Potency/Content of Protein Products

Clearly, the strength, or potency, of any pharmaceutical product is an important parameter which must be routine monitored. In the case of small molecules, strength (or potency) can be ascertained directly by physicochemical methods (e.g., liquid chromatography). However, for proteins the content, as measured by physicochemical methods, and biological potency are not necessarily equivalent.

Historically, the potency of therapeutic protein products has been determined using in-vivo or in-vitro bioassays selected to monitor the particular biological activity of interest for each particular product. For example, insulin potency was established by determining its hypoglycemic effects in rabbits, and human growth hormone preparations were monitored by their growth-promoting effects in hypophysectomized rats. While such tests are rather expensive to conduct and imprecise, they were appropriate, and indeed necessary, for monitoring potency of naturally derived materials due to the complex nature of many of these preparations. However, proteins derived from rDNA technology are generally much purer and chemically/physically well-defined compared to naturally derived products, and traditional bioassays are often too imprecise to provide meaningful batch-to-batch control of recombinant products. To overcome this limitation, the approach used for most recombinant products has been to extensively evaluate physicochemical techniques (e.g., chromatography, electrophoresis), as well as in-vitro bioassays (e.g., immunoassay, receptor-binding assays), with regard to how well such tests correlated with traditional in-vivo bioassays. In most instances physicochemical or in-vitro techniques have been established which correlate well with in-vivo bioassays and offer greatly improved precision and rapidity, as well as lower assay cost. The various types of physicochemical assays and in-vitro bioassays which have been employed are listed in Table VIII.

The establishment of a physicochemical potency assay for biosynthetic human growth hormone provides an interesting example of an approach for validation of such assays. The hypophysectomized rat growth promoting assay used tradionally has a precision of approximately 15-20% RSD, which is clearly not adequate for controlling well-characterized products. In order to select a suitable assay to replace the in-vivo bioassay, the biopotencies of the various chemically related substances in the product were established, the data for which are summarized in Table IX.

Clearly, all of the related substances, with the exception of hGH dimer, exhibited full biopotency as compared to hGH monomer. Further investigation of hGH dimer demonstrated that traditional immunoassays could not distinguish it from hGH monomer, although an immunoradiometric (IRMA) assay using two monoclonal antibodies could distinguish the two forms (8). As expected, size exclusion HPLC was also able to distinguish monomeric and dimeric forms of hGH. Further studies revealed that the IRMA assay precision was approximately 5-6% RSD, whereas size exclusion HPLC exhibited a precision of 1-2% RSD. Therefore, size exclusion HPLC was selected as the most useful candidate for replacement for the in-vivo bioassay.

Table VIII. Physicochemical and In-Vitro Bioassays Used for Potency/Content Determination

Total Protein Content
    Quantitative amino acid composition
    UV absorbance (280 nm)
    Kjeldahl nitrogen assay
    Colorimetric protein assay (e.g., Folin-Lowry)
Chromatography
    Reversed-phase HPLC (e.g., insulin)
    Size exclusion HPLC (e.g., human growth hormone)
In-Vitro Bioassays
    Competitive binding immunoassay (e.g., RIA)
    Enzyme-linked immunosorbent assay (ELISA)
    Immunoradiometric assay (IRMA) -- dual monoclonal
        antibody assay
    Receptor binding assay
    Cell binding assay
    Receptor mimics (anti-idiotypic antibodies)
    Immunoelectrophoresis (rocket electrophoresis)

Table IX. Biopotency of hGH Related Substances

COMPONENT	BIOPOTENCY, IU/mg (body weight gain)
hGH monomer	2.7-3.0
hGH dimer (non-covalent)	0.6
(Asp 149) desamido	3.0
(Asp 149,152) di-desamido	2.3
(Met 14) sulfoxide	2.3
Isolated total related substances mixture	2.8

The correlation between size exclusion HPLC and rat bioassay was extensively studied (9). The correlation of HPLC with the bio-assay was better than the correlation between the two bioassays (as judged from the correlation coefficients of the linear regression curves). These data, as well as the fact that no sign-ificant change in biopotency is observed under severe storage conditions (9), demonstrate that size exclusion HPLC is a suitable method to use in place of the hypox rat bioassay for routine batch-to-batch control of product potency. It is important to recognize that, in general, a variety of physicochemical tests, as well as a semiquantitative "bioidentity test" using the tradit-ionally bioassay, should be used for the characterization of each production lot to insure consistent product quality.

Extensive biopotency data for insulin related substances have also been obtained (10). As shown in Table X, dimeric forms of insulin are essentially inactive, whereas the A-21 desamido deri-vative is 90% as active as intact insulin.

Summary of Important Issues and Future Directions

The most important issues with regard to analytical technology/ strategy for rDNA derived products are summarized in Table XI. Some important future directions envisioned for this area of research/development are outlined in Table XII.

Table X. Biological Activity of Human Insulin Derivatives

Human Insulin Derivative	Bioactivity as % of Reference Standard
A21 Desamido insulin	93
B30 Des-threonine insulin	109
Arg AO insulin	70
Covalent insulin dimers	14
B3 Desamido insulin	98
B3 (beta-aspartyl)desamido insulin	106
A6-B7,A7-A11 insulin disulfide isomer	32
A11-B7,A6-A7 insulin disulfide isomer	36

Table XI. Important Issues Regarding Analytical
           Characterization of Recombinant Proteins

BIOPOTENCY
        -Assay precision/speed/cost
        -Correlation of in-vivo assays with alternative assays

PURITY ASSAYS
        -Assay time
        -Purity definition
        -Identification of component structures

CARBOHYDRATE CHARACTERIZATION
        -Structural detail routinely required
        -Correlation of structure and bioactivity

RESIDUAL DNA ANALYSIS
        -Assay time/ruggedness
        -Total DNA versus specific sequences

IDENTITY ASSAYS
        -More rapid/definitive tests

TERTIARY STRUCTURE (routine test methodology)

Development of analytical techniques for control of rDNA
derived products remains an area of exciting new directions and
challenges. Techniques to address the major analytical questions
are being developed at a rapid rate, and as experience is gained
through the successful commercialization of rDNA products, even
better analytical methodologies are expected to become available.

Table XII. Future Directions in the Analytical
            Characterization of Recombinant Proteins

BIOPOTENCY
        -Increased emphasis on receptor assays
        -Receptor mimics
PURITY ASSAYS
        -Routine use of capillary zone electophoresis (CZE)
        -Rapid HPLC (nonporous/small diameter porous particles)
IDENTITY TESTS
        -Peptide mapping by "rapid" HPLC
        -Peptide mapping by CZE
        -FAB/MS;PD/MS (coupled to microbore HPLC or CZE)
CARBOHYDRATE CHARACTERIZATION
        -Routine compositional/attachment site determination
        -Structure/functional activity relationships
        -CHO sequence by MS; NMR
RESIDUAL DNA DETERMINATION
        -Quantitative instrumental methods
        -General DNA detection

Literature Cited

1. Grossman,P.D.;Colburn,J.C.;Lauer,H.H.;Nielsen,R.G.;Riggin,R.M.;
    Sittampalam,G.S.; Rickard,E.C. Anal. Chem. 1989, 61, 1186-1194.
2. Riggin,R.M.,Dorulla,G.K.;Miner,D.J. Anal. Biochem. 1987, 167,
    199-209.
3. Kohr,W.J.;Keck,R.;Harkins,R.N. Anal. Biochem. 1982, 122,348-359.
4. Becker,G.W.;Tackitt,P.M.;Bromer,W.W.;LeFeber,D.S.;Riggin,R.M.
    Biotechnology and Applied Biochemistry 1988, 10,326-337.
5. Farid,N.A.;Atkins,L.M.;Becker,G.W.;Dinner,A.;Heiney,R.E.;
    Miner,D.J.;Riggin,R.M. J. Pharmaceutical and Biomed. Analysis
    1989,7,185-188.
6. Baker,R.S.;Schmidtke,J.R.;Ross,J.W.;Smith,W.C. Lancet 1981,
    2,1139-1141.
7. Farid,N.A.;Dydo,K.L. unpublished data.
8. Becker,G.W.;Bowsher,R.R.,Mackellar,W.C.;Poor,M.L.;Tackitt,P.M.;
    Riggin,R.M. Biotechnology and Applied Biochemistry 1987, 9,
    478-487.
9. Riggin,R.M.;Shaar,C.J.;Dorulla,G.K.;LeFeber,D.S.Miner,D.J.
    J.Chrom. 1988. 435, 307-318.
10. Chance,R.E. personal communication.

RECEIVED January 8, 1990

Chapter 7

Improvement and Experimental Validation of Protein Impurity Immunoassays for Recombinant DNA Products

Vincent Anicetti

Assay Services Department, Genetech, Inc., 460 Point San Bruno Boulevard, South San Francisco, CA 94080

The presence of impurities in protein pharmaceuticals may effect the biochemical quality of the product or result in detrimental effects on the drug recipient. To optimize separation processes for the removal of impurities highly sensitive assays are required. An advantage of recombinant DNA technology is that the production process and raw materials can be carefully controlled, which allows the development of highly specialized analytical methods. Antigen Selected Immunoassays (ASIA) designed to detect process specific proteins provide part-per-million sensitivity and can differentiate impurities from the product despite similarities in their physical-chemical properties. In this paper the general strategies for the design of these methods are reviewed and new studies exploring selected, critical aspects of these assays are also reported. These areas include the use of immunoaffinity chromatography as a method of enhancing the antibody response to poorly immunogenic E. coli proteins, the selection and characterization of E. coli protein reference materials and improvements in assay sensitivity. These results show that immunoaffinity chromatography, used together with two dimensional gel electrophoresis, can provide significant gains in the production and characterization of these complex reagents and provide the potential for further improvements in product purity.

0097–6156/90/0434–0127$06.00/0
© 1990 American Chemical Society

128   ANALYTICAL BIOTECHNOLOGY

The immunogenicity often associated with protein pharmaceuticals
intended for long term and frequent administration has emphasized the
need for high purity in these products. Not until the advent of recombinant
DNA technology, however, could proteins such as insulin (1) or human
growth hormone (2) be produced from well characterized and controlled raw
materials. The routine availability of large amounts of a protein product in
reproducible fermentation harvest materials has allowed the development
of highly refined purification processes and analytical methods. The
combination of these systems and analytical methods have resulted in the
production of gram amounts of proteins at purity levels exceeding 99.99%.

These purity levels are required because protein impurities have the
potential to effect the biochemical quality of the drug or result in
detrimental effects to the recipient, such as the initiation of an immune
response. The immunogenicity associated with human growth hormone is
generally believed to be a multicomponent phenomenon. The individual
patients' immune status (2,3), aggregation or denaturation of the protein
(3,4), additional amino acids (2-5) and the presence of host cell (E. coli)
components (5,6) are potential contributing factors. Current versions of
human growth hormone are quite pure (5,6) and induce a relatively low
incidence of anti-hGH production (7). The production of antibodies to
process-specific E. coli proteins has not been observed with currently
manufactured materials (2). Direct correlations between product purity and
adverse reactions have been made previously, however. Early preparations
of growth hormone used for investigative studies were pyrogenic in one
instance (5) and highly immunogenic in another (3). Both preparations
contained significant levels of E. coli components and the reduction of these
levels was directly correlated with either the absence of pyrogenicity or a
decrease in antibody production (3,5).

These findings emphasize the strong relationship between drug purity
and safety. They have further underscored the need for sensitive and
objective methods for evaluating host cell impurities in recombinant DNA
drug products. This evaluation serves both as an indicator of drug purity
and as a measure of consistency in drug manufacturing (8,9). While
traditional methods of protein purity analysis such as silver-stained
SDS-PAGE (10,11) or chromatographic techniques (9) are sensitive and
widely accepted, these methods will generally fail to detect impurities which
co-migrate or co-elute with the product. Further, such methods frequently
suffer the limitation that they are variations on the purification process
steps and thus cyclic in their analyses. A solution to this problem resulted
from the use of immunoassays for the detection of E. coli protein
contaminants (6,12,13). These systems are advantageous because the
specific nature of antibody-antigen binding allows the detection of
impurities in the presence of a large excess of the protein product and
regardless of most physiochemical similarities between the contaminating
protein(s) and the product.

These immunoassays rely on the production of polyclonal antibodies to
host cell and medium proteins which represent the most probable final

product impurities. These reference protein impurities are used as the standard material in a typical sandwich ELISA and the final product is measured against this reference material (6). The key steps in the development of these assays is shown in Figure 1. Immunoassays normally detect ng/mL amounts of protein, thus when a mg/mL or more of product is tested the assay system has a sensitivity at the part-per-million level (ng impurity per mg product). Although this design is simple in theory a number of technical problems reside within the development of these assays. These problems revolve around the selection of the protein reference material, the method for production of antibodies to the reference material and validation of the assay.

The production and isolation of the reference impurities will ultimately determine the validity of the assay. The reference material will define which impurities are detected because it is used to produce the antibodies used in for the assay. Also, because it is the standard against which the final product impurities are quantitated the distribution of proteins in the reference material should closely approximate that of final product material.

Based on total DNA content the E. coli genome could code for 3000 to 4000 individual proteins (14), and a total of perhaps 1500 have been visualized by two-dimensional electrophoresis during different growth conditions (15,16). Further, the proteins produced vary with the growth phase of the cell and the composition of the growth medium (6,15,16). These numbers are larger and the potential distribution more complex for mammalian host cell systems. From this complex mixture a process specific subset will be enriched, and therefore the distribution reference impurities must be refined to represent that population.

The process specific selection of the host cell proteins is an absolute requirement because the sensitivity and specificity of the assay system is determined largely by the number of proteins in the reference material. This is a result of the mechanical limitations of ELISA systems and the immune response, which will be biased to those proteins in the highest concentration or which are the most immunogenic. Our studies of impurities isolated at various steps in the growth hormone process revealed that most components observable by two dimensional gel electrophoresis and silver staining at any particular process step were not observable by the same analysis at the preceding process step (V. Anicetti, unpublished data). The isolation of the reference impurities at the fermentation harvest or too early in the purification process will result in an assay directed to largely irrelevant proteins. These and similar observations (6) suggest that a generic impurity assay will not be possible without a major change or improvement in the assay technology.

The ideal reference preparation would contain exactly those host cell proteins, medium proteins and process raw materials (i.e., a monoclonal antibody) that are present in any particular final product lot. Further, the reference impurities would be present in the same distribution and biochemical quality as those in the final product. Without knowledge of the

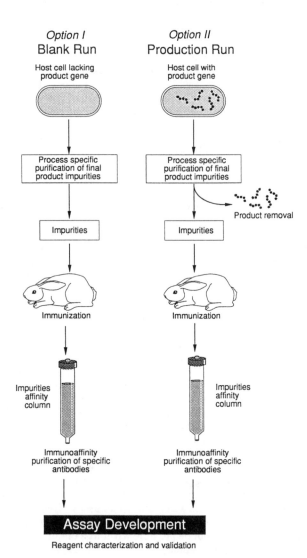

Figure 1.   Key steps in the development of protein impurity assays. The
reference impurities may be obtained by a process specific
purification of host cell proteins arising from a blank run or a
production run. While the production run is a more accurate
population of potential impurities, the product removal step
involves significant technical difficulty.

identity of impurities at the part-per-million level and because most
purification processes will have subtle variation, the selection of the
reference material can, at best, only approximate the most likely final
product impurities (6,17). Further, a prudent approach suggests the
reference preparation be broader in composition than the normal final
product population to allow for small variations in fermentation and
purification from run to run. It is generally agreed that the best
representation of the manufacturing process results from a large scale
fermentation and purification run in which broader pools are taken to
anticipate minor variations in individual chromatographic behavior.

Two general approaches to production of the reference impurities may
be used and either is acceptable with the appropriate validation studies.
The first approach attempts the isolation of impurities at a suitable step in
the purification process during a manufacturing run. This isolation is
followed by a further separation of the product from the impurities, which is
normally achieved by immunoabsorption.

This strategy is attractive because the impurities will reflect exactly
those produced by the host cell during product expression and which co-
purify with the product through the process. The immunoabsorption
procedures required for this approach, however, are difficult to perform and
validate at the part-per-million level (i.e., ng of impurity per mg of protein
product). All of the product, product fragments and product-impurity
complexes must be removed without the loss of any of the impurities. Any
residual product will produce antibodies during immunization and render
the assay nonspecific. Conversely, the immunoabsorbtion step must not
remove any of the impurities. An exact demonstration of these criteria at
the ppm level is not possible with currently available non-antibody based
analytical methods.

An alternative is the 'blank run' approach where a fermentation of the
host cell lacking the product gene is performed and purified through the
process to a point where the product is normally 95 to 99% pure (6,17). Two
key assumptions are made in this strategy, the first is that the lack of
product expression does not significantly effect the population of in-process
impurities and second, that the lack of product does not significantly effect
the chromatography of the impurities. To test the latter, an E. coli paste
was purified through the growth hormone process in the presence and
absence of a small amount of growth hormone and analyzed by two
dimensional electrophoresis with silver staining (Figure 2).

Examination of the resulting gels demonstrated the reference
impurity population to be quite complex, composed of over one hundred
individual spots. It was also apparent that the two purification runs, which
were performed at manufacturing scale, demonstrated a remarkably similar
distribution of impurities. Finally, the addition of a small but significant
amount of growth hormone (approximately 5-8%) to the starting cell extract
did not appear to grossly affect the distribution of the impurities and, thus,
their chromatographic behavior in the process (Figure 2, panel A). The
exact experiment, where growth hormone is 95 to 99% of the protein load,

A   B

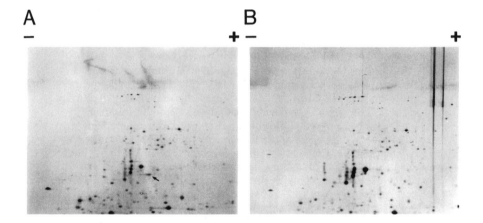

Figure 2.   Comparison of 'blank run' reference impurities obtained from a
             single E. coli lysate but purified through the process in the
             presence of growth hormone (panel A, arrow) or the absence of
             growth hormone (panel B). These silver stained 2-D gels
             demonstrate that the absence of the product did not
             significantly change the distribution of impurities and therefore
             their chromatographic behavior in the process.

cannot be performed because of the sensitivity limitations of silver stain for any individual component. These data, however, provide an adequate level of assurance that the 'blank run' approach is reasonable for this process.

The detection and accurate quantitation of any protein in an immunoassay requires that a condition of antibody excess exist. This is required for each protein in the reference impurity preparation. The acquisition and characterization of broad spectrum antisera against complex protein mixtures, therefore, is a fundamental goal in the development of these assay systems.

The production of broad spectrum antisera to protein mixtures presents two problems. First, immunization with mixtures of proteins may be complicated by their distribution in the mixture and the properties of each component. Also, the host response may be dominated by antigens in high concentration or those which are strongly immunogenic. While the evolutionary distance of E. coli from mammals might suggest that all E. coli proteins would be strong immunogens, examples of shared antigenic determinants are known and it is possible that tolerance to these determinants may occur (18-20). Tolerance is a greater concern with mammalian cell production systems where production may be in rodent cells (i.e. Chinese Hamster Ovary cells) and the immunization program is conducted in a closely related species (rabbit or mouse). Finally, antigenic competition has been demonstrated as a complicating factor in immunization procedures using mixtures of E. coli proteins (20,21).

Second, methods for the characterization of complex antisera are difficult. Antisera to E. coli protein mixtures have been developed with impressive spectra of reactivity using conventional immunization methods (6,22-23). An exact assessment of the spectrum of antibody reactivity is often limited, however, by the resolution of the analytical methods used. Counter immunoelectrophoresis is limited by the relatively low sensitivity of detection and resolution for complex mixtures of reacting species. One dimensional silver stained SDS-PAGE and immunoblotting provides sensitive detection limits but lacks resolution. Therefore, methods which have a high degree of resolution and sensitivity are required to best compare potential improvements in the production of antibodies to minor components in the mixture.

Two immunization procedures designed to enhance the immune response to multiple antigen mixtures have been reported recently. The cascade immunization technique (20) utilized in vitro depletion of E. coli proteins (ECPs) which had previously elicited an antibody response. The removal of these dominant immunogens from the mixture was accomplished by immunoabsorption with antibodies obtained from an earlier antiserum. The passive immunization procedure (21) relied on in vivo blocking of strong immunogens by the concurrent administration of early antiserum obtained previously. This latter report demonstrated the presence of an apparently poorly immunogenic ECP to which a humoral response could only be elicited by this passive procedure.

134 ANALYTICAL BIOTECHNOLOGY

Given the potential of these methods to improve antibody production to multiple antigen mixtures, we performed a comparison of these two methods to a conventional immunization procedure reported previously (6). In addition, we selected two dimensional electrophoresis and immunoblotting as a tool to examine the production of antibodies to minor components of the ECP mixture.

For these studies a cell paste of E. coli K-12 containing the plasmid pBR322 was processed by ammonium sulfate precipitation and ion exchange chromatography to mimic typical purification process steps. The isolated proteins were then used as the immunogen for three groups of three rabbits each.

The first group underwent a conventional immunization procedure (6) in which one half mg of ECPs were administered subcutaneously in Complete Freunds Adjuvant (CFA) on day one and in Incomplete Freunds Adjuvant (ICFA) on day 7. The rabbits were boosted every 14 days for 160 days and serum collected 7 days after each boost.

A second group underwent the passive immunization procedure. The protocol was similar to the conventional immunization procedure except injections of antigen also included a concurrent intravenous injection of 0.5 ml of serum obtained from the same individual rabbit seven days earlier.

The third group underwent a modification of the cascade immunization procedure (24). After a primary injection of ECP in CFA and a subsequent injection in ICFA a serum sample was taken seven days later and the IgG fraction used to prepare an affinity column. The entire ECP mixture was passed over the column and fractions which were depleted of one or more ECPs (as determined by silver stain SDS-PAGE and compared to the starting preparation) were used for the subsequent immunization injection. This procedure of ECP adsorption was repeated with subsequent antisera obtained on days 28 and 42 and the resulting depleted ECP fractions used for injections on either days 35 or 49 and 62, respectively (Figure 3). Thereafter the rabbits received injections as described in the normal immunization procedure.

It was clear from this study that affinity chromatography with early antibodies (after careful analysis and pooling of the column flowthrough by a very sensitive method such as silver stain SDS-PAGE) can be used to alter the distribution of the reference immpurity protein mixture toward those components which were in low concentration or poor immunnogens.

The best method to determine the success of antibody production to the minor components was made by two dimensional SDS-PAGE and immunoblotting (24). A comparison of the antisera (day 112 antisera) from the three groups demonstrated that the cascade immunization antisera detected a number of minor components (Figure 4C, arrows) which were not observed with the conventional or passive antisera (Figure 4B and D). It was clear from these results that the cascade antisera was far superior in its spectrum of antibody reactivity and, in fact, was comparable or superior in detection of ECPs to silver stain (Figure 4A). Although silver stain appeared to have an improved detection of certain low MW or basic proteins,

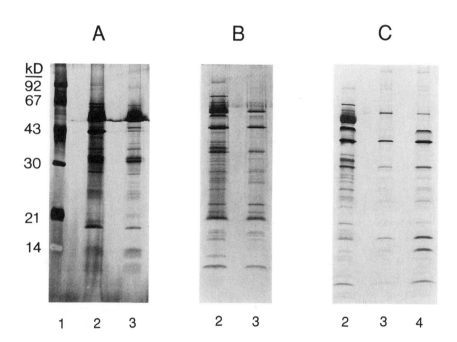

Figure 3.   Selection of ECP subpopulations for progressive iterations of the cascade procedure by silver stained SDS-PAGE. Lane 2 in each panel shows the entire ECP mixture used as the column load and lane 3 shows the column flowthrough fraction used for the next injection. Panel A demonstrates the affinity chromatography performed with day 14 antisera, Panel B with day 28 antisera and Panel C with day 42 antisera. The arrow shows ECPs depleted by the early antibodies. The progression of the immune response is clearly apparent although it is clear not all of these proteins are equally immunogenic. A 50 Kd protein has saturated its respective antibody and begun to flow through the column (Panel C, lane 4). Reproduced with permission from Ref. 24. Copyright 1989 The Humana Press Inc.

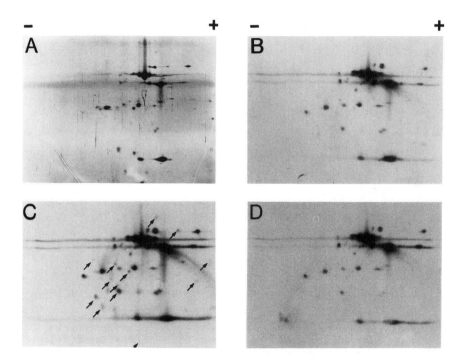

Figure 4.   Separation and detection of ECPs by two dimensional gel
electrophoresis. A: Silver stained. B: Immunoblot with
conventional procedure day 112 antisera. C: Immuunoblot with
cascade procedure day 112 antisera. D: Immunoblot with
passive immunization procedure day 112 antisera. Exposure
time was 24 hours. Reproduced with permission from Ref. 24.
Copyright 1989 The Humana Press Inc.

a longer exposure of the blots to increase sensitivity showed that a large number of additional spots could be detected. These included many of the low MW and basic components observed only by silver stain previously and may have reflected the lesser immunogenicity or differential transfer of these proteins. Many spots which were detected only by the cascade antisera at 24 hours could be detected by the conventional antisera at 48 hours. This suggested that the differences in reactivity were primarily the result of differences in antibody titer.

Different reactivity of antisera were not the result of differential transfer of proteins because the same sheet of nitrocellulose was used for all of the blots after washing with acid between experiments. The addition of ^{125}I labeled protein A without primary antisera to the acid washed sheet did not reveal the presence of non-acid elutable antibodies after a 24 hour exposure. Also, proteins were not eluted from the sheet because re-blotting with the cascade antisera demonstrated the identical pattern after blotting had been performed with each of the group antisera (24).

These data suggested that a mechanism of early priming of the immune response though the cascade procedure resulted in a broader spectrum of antibody reactivity. This improvement also required additional time (56 days) and/or subsequent injections of the total antigen mixture because similar experiments with day 56 antisera demonstrated equivalent antisera reactivity (24).

The continuing refinement in the selection of reference materials and the production of antibodies to complex protein mixtures has resulted in immunoassay systems of remarkable sensitivity and specificity. In particular, the selection and enrichment of the antibody population by immunoaffinity purification against the reference impurities has afforded an additional level of control over the production and validation of these reagents and served to improve the assay range and sensitivity (6,17). This normalization of the antibody population to a stoichiometric relationship with the reference impurities has suggested the term Antigen Selected Immunoassay (ASIA) for these methods.

The improvement of ASIA methodology and validity has resulted in a continued refinement of purification processes and improved product quality. With each incremental improvement in product purity, however, any further advance is at least partially limited by the sensitivity of the assay system. Thus, improvement of product quality is often directly linked to improved sensitivity in the corresponding analytical method. One method commonly employed to improve immunoassay sensitivity is through the use of $F(ab)_2$ antibody fragments (25). Using an ECP assay for growth hormone, we examined the use of $F(ab)_2$ antibodies as a reagent in comparison to the intact IgG antibody (Figure 5). It was clear from this study that the $F(ab)_2$ reagent resulted in a far more sensitive and precise assay system. This improvement was achieved primarily by decreasing the non-specific noise associated with this system. The resulting detection limit of 50 pg/mL ECP corresponds to a sensitivity of 0.02 ppm when 2.5 mg/mL hGH is tested (our normal assay concentration). Thus the careful

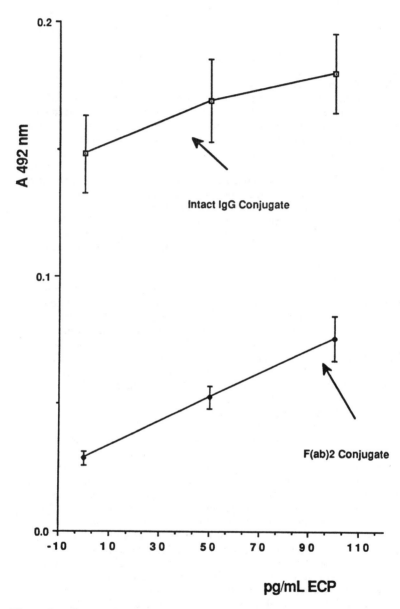

Figure 5.   Comparison of the sensitivity of immunoassay systems for the
             detection of ECP in hGH. Each point shown represents the
             mean response of twenty replicate determinations.

production, selection and refinement of antibodies can provide the ability to detect protein impurities approaching the one ppb level.

Conclusion
The validity of any statement about the purity of a protein is directly linked to the quality of the analytical method used. The validation of immunoassay systems to detect protein impurities in rDNA pharmaceuticals must be achieved by careful production and characterization of the assay reagents. The studies presented here demonstrate that the blank run approach is reasonable for the isolation of reference materials and that high quality broad spectrum antisera can be produced to these mixtures. Significant improvements in assay sensitivity approaching the ppb level are attainable and should provide the methods to further improve product purity.

Acknowledgments
The author would like to thank Martin Simonetti, Andrew Jones, Linda Blackwood, Ken Olsen and Anthony Chen for their expert assistance and advice. Special thanks to the Genentech Document Processing staff for the preparation of this manuscript.

Literature Cited
1. Keen, H.; Glynne, A.; Pickup, J.C.; Viberti, G.C.; Bilous, R.W.; Jarrett, R.I.; Marsden, R. Lancet II 1980, 398-401.
2. Kaplan, S.L.; Underwood, L.E.; August, G.P.; Bell, J.J.; Blethen, S.L.; Blizzard, R.M.; Brown, D.R.; Foley, T.P.; Hintz, R.L.; Hopwood, N.J.; Johanson, A.; Kirkland, R.T.; Plotnick, L.P.; Rosenfeld, R.G.; Van Wyk, J.J. Lancet I 1986, 697-700.
3. Fryklund, L.; Brandt, J.; Hagerman, M.; Pvlu, B.; Skoog, B.; Wichman, A. In Human growth hormone; Raiti, S.; Tolman, R.A., Eds.; Plenum Medical Book Co: New York, 1986, p 257.
4. Moore, W.V.; Leppert, P. J. Clin. Endocrin. Metabol. 1980, 51(4), 691-97.
5. Ross, M.J.; Olson, K.C.; Geier, M.D.; O'Connor, J.V.; Jones, A.J.S. In Human growth hormone; Raiti, S.; Tolman, R.A., Eds.; Plenum Publishing Corp: New York, 1986, p 241.
6. Anicetti, V.R.; Fehskens, E.F.; Reed, B.R.; Chen, A.B.; Moore, P.; Geier, M.D.; Jones, A.J.S. J. Immunol. Meth. 1986, 91, 213-24.
7. Genentech, Inc. National cooperative growth study. Summary report 5. January 1987, Genentech, Inc. 460 Pt. San Bruno Blvd, South San Francisco, CA 94080.
8. Jones, A.J.S.; O'Connor, J.V. In Developments in biological standardization; Perkins, F.T.; Hennessen, W., Eds.; S. Karger: Basel 1985, vol. 59, p 175.
9. Hancock, W.S. Chromatography Forum 1986, 1(3), 57-9.
10. Oakley, B.R.; Kirsch, D.R.; Morris, N.R. Anal. Biochem. 1980, 105, 361-3.
11. Morrissey, J.H. Anal. Biochem. 1981, 117, 307-10.
12. Baker, R.S.; Schmidtke, J.R.; Ross, J.W.; Smith, W.C. Lancet II, 1981, 1139-92.

13. Ross, J.W.; Baker, R.S.; Hooker, C.S.; Johnson, I.S.; Schmidtke, J.R.; Smith, W.C. In Hormone drugs; United States Pharmacopeial Convention, Inc. Rockville: 1982, p 127.

14. Szekely, M. From DNA to protein. The transfer of genetic information; John Wiley and Sons: New York, 1980, p 13.

15. Neidhardt, F.C.; Vaughn, V.; Phillips, T.A.; Block, P.L. Microbiol. Rev. 1983, 47(2), 231-84.

16. O'Farrell, P.H. J. Biol. Chem. 1975, 350(10), 4007-21.

17. Jones, A.J.S. In The Impact of Chemistry on Biotechnology; Phillips, M.; Shoemaker, S.P.; Middlekauff, R.D.; Ottenbrite, R.M., Eds.; American Chemical Society: Washington, D.C., 1988, p 193.

18. LeRoith, D.; Shiloach, J.; Roth, J.; Lesniak, M.A. J. Biol. Chem. 1981, 256, 6533-6.

19. Soderstrom, T.; Hansson, G.; Larson, G. N. Engl. J. Med. 1984, 310, 726-7.

20. Maruo, T.; Cohen, H.; Segal, S.J.; Koide, S.S. Proc. Natl. Acad. Sci. USA 1979, 76, 6622-6.

20. Thalhamer, J.; Fruend, J. J. Immunol. Meth. 1984, 66, 245-51.

21. Thalhamer, J.; Freund, J. J. Immunol. Meth. 1985, 80, 7-13.

22. Gooding, R.P.; Bristow, A.F. J. Pharm. Pharmacol. 1985, 37, 781-6.

23. Owen, P. In Electroimmunochemical analysis of membrane proteins; Bjerrum, O.J., Ed.; Elsevier: Amsterdam, 1983, p 348.

24. Anicetti, V.R.; Simonetti, M.A.; Blackwood, L.L.; Jones, A.J.S.; Chen, A.B. Appl. Biochem. Biotech. 1989, 22(2), 1-10.

25. Ishikawa, E.; Imagawa, M.; Hashida, S.: Yoshitake, S.; Hamaguchi, Y.; Uneo, T. J. Immunoassay 1983, 4(3), 209-327.

RECEIVED December 20, 1989

Chapter 8

Minimum Variance Purity Control of Preparative Chromatography with Simultaneous Optimization of Yield

An On-Line Species-Specific Detector

Douglas D. Frey

Department of Chemical Engineering, Yale University, New Haven, CT 06520

Statistical process control methods are applied to preparative chromatography for the case where cut points for the effluent fractions are determined by on-line species-specific detection (e.g., analytical chromatography). A simple, practical method is developed to maximize the yield of a desired component while maintaining a required level of product purity in the presence of measurement error and external disturbances. Relations are developed for determining tuning parameters such as the regulatory system gain.

Several investigators [see, e.g., Kalghatgi and Horváth (1) and DiCesare et al. (2)] have developed columns and instrumentation for carrying out rapid analytical liquid chromatography which are well suited for on-line monitoring in the biotechnology and pharmaceutical industries. One potential application for on-line monitoring using analytical chromatography is as a species-specific detector for determining effluent cut points in preparative chromatography. A detection system of this type has the potential to meet product quality and system performance requirements for a preparative chromatographic separation more reliably than if simple timers or nonspecific detectors are used.

Figure 1 illustrates a preparative chromatographic process which employs on-line analytical chromatography as a species-specific detector. In general, the solute to be purified (i.e., the desired component) is produced either continuously or in batches in an upstream process and usually enters an intermediate storage tank before it is introduced in individual feed slugs into the preparative column. As shown in the figure, a number of streams are fed at programmed times to the inlet of the preparative column. The compositions of these inlet streams depend on the particular type of preparative chromatography under consideration, as discussed below. The effluent from the preparative column is monitored by the analytical chromatographic system. The

0097–6156/90/0434–0141$06.00/0

information derived from this monitoring is then used to actuate a valve which directs effluent fractions either into a product tank, into a tank containing material to be recycled, or into a waste tank. The control of the valve which sends the column effluent into collection tanks (i.e., the location of the effluent cut points) is a key factor in accomplishing a number of objectives, such as optimizing the recovery of the desired component or attaining a required product purity.

Figure 2 illustrates the effluent concentration profile for a particular operating mode of preparative chromatography, termed displacement development, in which a displacer solution is used to separate a desired component (component 2) from two impurities (components 1 and 3). This type of preparative chromatography involves first equilibrating the column with a carrier and then introducing a feed slug into the column which contains the desired component together with the impurities. Next, a displacer solution containing a component which is more strongly adsorbed than any of the other components is introduced into the column such that a succession of contiguous bands is produced in the column effluent. After the desired component is collected in the product fraction, the displacer solution is washed from the column, the column is cleaned, and then re-equilibrated with the carrier so that the cycle can be repeated. Further details concerning the operation of displacement chromatography are discussed in reference 3.

Also shown in Figure 2 are the locations of two cut points defined by the instantaneous purities P_u and P_d, where P is the mass fraction of desired component. It should be noted that on-line species-specific detection for determining cut points is particularly useful in displacement chromatography since the formation of bands in close proximity to each other makes nonspecific detection (e.g., the monitoring of UV absorbance) ineffective. Note also that fixed values for P_u and P_d will in general not correspond to a fixed value for the purity of the product fraction, which is presumed to be measured off-line and will be denoted by the symbol P_p. For example, if the concentration of the desired component in the feed tank to the preparative column is reduced or if the relative sensitivity of the monitoring method to the impurities decreases, the values of P_u and P_d used to define cut points would have to be increased if the desired average purity for the entire product fraction is to be maintained.

Objectives

In addition to the actual analytical instrumentation and associated equipment shown in Figure 1, another aspect of on-line monitoring is the use of information acquired from these monitors to fullest extent possible. In particular, although using specific detectors to determine cut points in preparative chromatography has advantages over using simple timers, preparative chromatographic processes employing specific detectors are still subject to measurement error (e.g., run-to-run variations due to baseline drift in the analytical chromatographic system) as well as to uncontrollable external

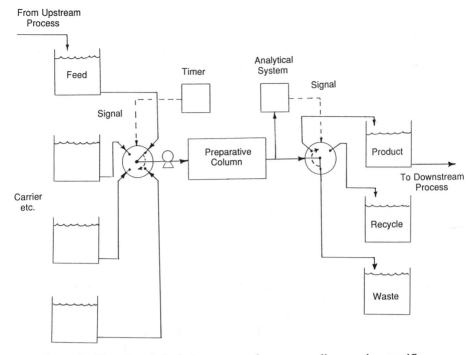

Figure 1. Use of analytical chromatography as an on-line species-specific detector for a preparative chromatographic process.

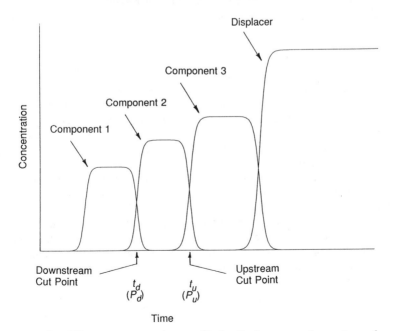

Figure 2. Effluent concentration profile in displacement chromatography.

disturbances, such as changes in feed composition or imprecisely repeated column cleaning procedures between successive production cycles.

The purpose of this study is to develop a simple method which can be applied to preparative chromatography for the purpose of maximizing the yield of the desired component (i.e., maximizing the fractional recovery of desired component in the product fraction) while achieving a required product purity in the presence of measurement error and external disturbances. The development presented in this study applies to all modes of preparative chromatography (e.g., gradient elution, displacement development etc.) where the column is overloaded with feed such that bands for two impurities (one located upstream and one downstream of the desired component) are not completely resolved from the band for the desired component. For simplicity, this study will be restricted to the case where the desired component is collected in a single product fraction from the column effluent and where recycle of partially purified product is not employed. Furthermore, it is assumed that a tight constraint is imposed on purity; i.e., purity is not allowed to vary in either direction from an optimal set point. The value of the purity set point is generally chosen to address two considerations: (i) purities lower than a certain minimum purity require that the entire product fraction be reprocessed and (ii) purities higher than needed for subsequent downstream processing result in a lower yield than could be attained otherwise. Specifically, this study will discuss the design and operation of an on-line feedback system which uses statistical methods (see references 4-8) in order to perform the following tasks: (i) identification of statistically significant trends in product purity in the presence of measurement error, (ii) correction of cut point locations as defined by species-specific on-line detection before product fractions deviate significantly from a purity specification, and (iii) optimization of product yield.

Process Variables

The variables P_d, P_u, t_u, t_d, P_p, and Y will be defined only at time points corresponding to when P_p is measured; i.e., these variables will be discrete in time rather than continuous and will incorporate the subscript k to denote a specific time point corresponding to each production cycle. In addition, the operator Δ will be used to denote that a variable is evaluated as the deviation between its actual value and a reference value corresponding to a cyclic steady state; i.e., $\Delta P_{p,k}$ is the deviation variable for product purity for the time point k and is equal to the actual product purity minus the desired (i.e., set point) purity. It is assumed in this study that measurement of the output variable ΔP_p at the time point k occurs simultaneously with the adjustments to the input variables ΔP_u and ΔP_d which affect the product fraction produced at the time point $k+1$. In addition, since recycle streams are not considered in this study, it follows that if $\Delta P_{u,k}$ differs from $\Delta P_{u,k-1}$, the full effect of this change is manifested in a change in ΔP_p from its value for the time point k to its value for time point $k+1$.

Optimization of Yield

One advantage in using specific detection to monitor the effluent from a preparative chromatographic process is that it becomes a simple matter to maximize the yield of the desired component for a fixed product purity. In particular, if A_i and $A_{i,feed}$ are the amounts of component i in the product fraction and feed slug, respectively, and if component 2 is the desired component, then P_p, P_u, P_d, Y, and dA_i are given as follows (see Figure 2):

$$P_p = \frac{A_2}{A_1 + A_2 + A_3} \tag{1}$$

$$P_u = \frac{C_{2,u}}{C_{2,u} + C_{3,u}} \tag{2}$$

$$P_d = \frac{C_{2,d}}{C_{1,d} + C_{2,d}} \tag{3}$$

$$Y = \frac{A_2}{A_{2,feed}} \tag{4}$$

$$dA_i = F(C_{i,u} \, dt_u - C_{i,d} \, dt_d) \tag{5}$$

Equations 1 - 5 lead to the result

$$dP_p = \frac{1 - P_p}{A_1 + A_2 + A_3} \, dA_2 + \frac{F \, P_p}{A_1 + A_2 + A_3} \, (C_{1,d} \, dt_d - C_{3,u} \, dt_u) \tag{6}$$

Equation 6 indicates that if changes in the cut point locations (i.e., dt_u and dt_d) are chosen to satisfy the constraint $dP_p = 0$, the further restriction $P_u = P_d$ ensures that $dA_2 = 0$. This implies that the largest yield for a given product purity is attained simply by setting $P_u = P_d$. Thus, in the regulatory system developed in this study, the required product purity will be attained by manipulating a single independent input variable (i.e., P_d) and the yield will be maximized by setting $P_{u,k} = P_{d,k}$.

Linearized Process Model

For the case where perturbations from a cyclic steady are moderate in magnitude, it is convenient to employ a linearized process model when designing a regulatory system. In particular, if the effluent profiles are linearized locally about the upstream and downstream cut points, and if the yield is maximized by enforcing the restriction $P_{u,k} = P_{d,k}$, Equations 1 - 5 lead to the following result:

$$\Delta P_{p,k} = \left[\frac{\frac{dt_d}{dP_d}(P_p C_{1,d} - (1 - P_p)C_{2,d}) - \frac{dt_u}{dP_u}(P_p C_{3,u} - (1 - P_p)C_{2,u})}{A_1 + A_2 + A_3} \right] FB\Delta P_{d,k}$$

(7)

where the entire quantity in square brackets in Equation 7 is the steady-state process gain (which will be denoted by the symbol g) and is evaluated at an average value characteristic of the operating range of interest. Note that B is the backward shift operator defined by the relation $B^n \Delta P_{d,k} = \Delta P_{d,k-n}$.

Disturbance Model

One aspect of developing a statistical regulatory system for P_p is to identify the behavior of this variable if P_u and P_d are held constant. For the case of preparative chromatography, the resulting behavior of P_p would tend to be a combination of fluctuations with a fixed average (e.g., white noise measurement error from the on-line monitoring method) and fluctuations which drift in time (e.g., random walk behavior resulting from external disturbances, such as variations in feed composition). The net result is a disturbance model of the following form:

$$D_k = X_k + \beta_k$$

(8)

where D_k is the value of $\Delta P_{p,k}$ when P_u and P_d are held constant, X_t is a random walk process, and β_k is a white noise process with zero mean. Note that a random walk process has the property that the difference in its values from the k to the $k+1$ time point is described by white noise. X_k can therefore be written in the form

$$(1 - B) X_k = \alpha_k$$

(9)

where α_k is a second (independent) white noise process. Equations 1 and 2 can be combined to yield

$$(1 - B) \, D_k = (1 - \theta \, B) \, (\alpha_k + \beta_k) \tag{10}$$

which is the standard form for a first-order integrated moving-average process (4). In Equation 10, θ is a parameter varying between zero and unity which characterizes the relative importance of white noise measurement error. Methods for determining θ from process data are described in reference 4.

The minimum variance prediction of D_{k+1} using information at D_k is given by the quantity $D_k - \theta \, (\alpha_k + \beta_k)$. Since the error in this prediction is given by $\alpha_{k+1} + \beta_{k+1}$, it follows that when Equation 10 applies, the following relation also applies for the minimum variance prediction of D_{k+1} (4):

$$(1 - \theta \, B) \, \hat{D}_{k+1/k} = (1 - \theta) \, D_k \tag{11}$$

Figure 3 illustrates typical behavior for the product purity when P_d and P_u are held constant. This figure was calculated using Equation 10 with a white noise process having a standard deviation of 0.0075 together and with $\theta = 0.5$. The basic goal of a regulatory system in this case would be to correct for the random walk component of the purity error, i.e., the downward drift in purity between cycles 30 and 80. As noted above, drifting behavior of this type can be envisioned as being caused by a change in feed tank composition which is in turn caused by perturbations having random walk behavior located upstream from the preparative column. Note that a correction for random walk behavior would be accomplished by adjusting the purities defining cut point locations (i.e., P_d and P_u) and should be performed in such a manner that the regulatory system does not respond unduly to white noise.

Regulatory System

Figure 4 illustrates the operation of an internal model control system (5) designed to use P_d as a manipulated variable to minimize the variance of the purity error ΔP_p while optimizing Y. As shown in the figure, the effect of the change in P_d at the time point $k-1$ is subtracted from the measured output variable (i.e., the purity error) at the time point k in order to determine an estimate of ΔD_k; i.e.,

$$\Delta D_k \approx \Delta P_{p,k} - g \, B \, \Delta P_{u,k} \tag{12}$$

This value for ΔD_k is then filtered (after accounting for purity set point changes) using a minimum variance predictor (see Equation 11) to provide an estimate of ΔD_{k+1}. The optimal manipulation of P_d at the time point k is then

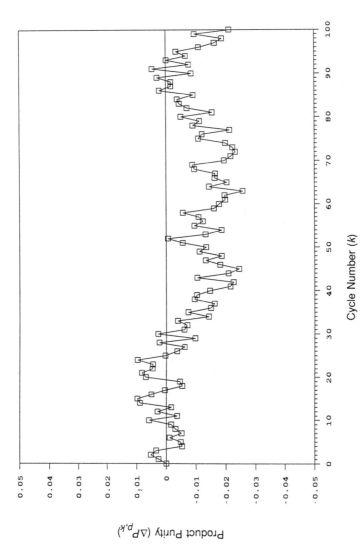

Figure 3. Simulated behavior of product purity for a preparative chromatographic process with P_u and P_d held constant. Calculations were performed using Equation 10 with $\theta = 0.5$.

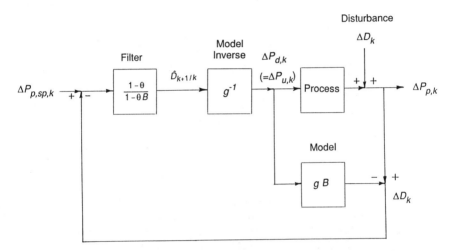

Figure 4. Internal model control representation of a minimum variance control method for product purity with simultaneous optimization of product yield.

determined using the inverse of the process model given by Equation 7; i.e.,

$$\Delta P_{u,k} = -g^{-1} \Delta \hat{D}_{k+1/k} \tag{13}$$

Although Figure 4 provides a conceptually simple framework for minimizing the variance of P_p from its set point, there is a much simpler relationship which can be used to relate the manipulated input ΔP_d to the purity error. In particular, Equations 9-12 can be combined to eliminate D_k, $\alpha_k + \beta_k$, and the minimum variance prediction of D_{k+1}. In addition, if g is evaluated using the process model described by Equation 7 and if the term $(1 - B)^{-1}$ is expanded to yield $1 + B + B^2 + ...$, then the following relation results:

$$\Delta P_{d,k} = - \left[\frac{\dfrac{dt_d}{dP_d}(P_p C_{1,d} - (1 - P_p)C_{2,d}) - \dfrac{dt_u}{dP_u}(P_p C_{3,u} - (1 - P_p)C_{2,u})}{A_1 + A_2 + A_3} \right]^{-1}$$
$$\times \frac{1-\theta}{F} \sum_{i=0}^{i=k} \Delta P_{p,i}$$

$$\tag{14}$$

Equation 14 describes a discrete integral controller which minimizes the variance of the purity error by summing the error in P_p for the time points from 0 to k, multiplying this sum by a controller gain, and then using the product to determine $\Delta P_{d,k}$. Note that values of θ near zero correspond to aggressive control which guards against random step changes in P_p caused by external disturbances for the case where there is little white noise measurement error. In contrast, values of θ near unity correspond to sluggish control, which is appropriate when white noise measurement error is dominant. Note also that in the absence of measured values for θ, this parameter can be employed as a semi-empirical tuning parameter which varies between zero and unity (8). In particular, an initial design would generally use a value of 0.5 for θ and, after operating experience is obtained for the particular system under consideration, θ can be adjusted either towards zero or unity to make the regulatory system either more or less tolerant of white noise.

Summary

Statistical process control methods are applied to preparative chromatography for the case where effluent cut points are determined by on-line species-specific detection. In particular, Equation 14 (with the quantity in square brackets evaluated at an average value characteristic of the operating range of interest) together with the restriction $P_{d,k} = P_{u,k}$ can be used as a simple, practical method for minimizing the variance of the purity error while optimizing the product yield in the presence of measurement error and external disturbances.

Acknowledgment

The author would like to thank Professor Csaba Horváth for helpful comments on this work.

Legend of Symbols

A	total amount of component in product fraction (g)
B	backward shift operator
C	concentration (g/l)
D	uncontrolled output (disturbance)
$\hat{D}_{k+1/k}$	minimum variance prediction of D_{k+1} from D_k
F	volume flow rate (cm^3/s)
P_d, P_u	purities defining downstream and upstream product cut points
P_p	purity for product fraction
t	time (s)
X	random walk process
Y	yield of desired component in product fraction

Subscripts

$1,3$	impurities
2	desired component
d	downstream cut point
feed	feed slug

k   cycle number

p   product fraction

sp   set point

u   upstream cut point

Greek Symbols

α, β white noise processes

Δ deviation from a steady state

θ parameter in Equation 10

Literature Cited

1. Kalghatgi, K.; Horváth, Cs. *J. Chromatogr.*, 1987, 398, 335.
2. DiCesare, J.L.; Dong, M.W.; Ettre, L.S. *Introduction to High-Speed Liquid Chromatography*, Perkin Elmer Corporation: Norwalk, CT, 1981.
3. Frenz J.; Horváth, Cs. "High Performance Displacement Chromatography," in *HPLC-Advances and Perspectives*, Vol. V, Cs. Horváth (editor), Academic Press: New York, pp. 211-314, 1989.
4. Astrom, K.J.; Wittenmark, B. *Computer Control Systems: Theory and Practice*, Prentice-Hall: Englewood Cliffs, NJ, 1984.
5. Box, G.E.P.; Jenkins, G.M. *Time Series Analysis: Forecasting and Control*, Holden Day: San Francisco, 1970.
6. Morari, M.; Zafiriou, E. *Robust Process Control*, Prentice Hall: Englewood Cliffs, NJ, 1989.
7. MacGregor, J.F. *Canadian J. Chem. Eng.*, 1973, 51, 468.
8. Palmor, Z.J.; Shinnar, R. *Ind. Eng. Chem. Process Des. Dev.*, 1985, 18(1), 427.

RECEIVED April 13, 1990

Chapter 9

Amino Acid Sequence–Mass Spectrometric Analyses of Mating Pheromones of the Ciliate *Euplotes raikovi*

R. A. Bradshaw[1], S. Raffioni[1,2], P. Luporini[2], B. Chait[3], T. Lee[4], and J. Shively[4]

[1]Department of Biological Chemistry, University of California, Irvine, CA 92717
[2]Department of Cell Biology, University of Camerino, Camerino, Italy
[3]The Rockefeller University, New York, NY 10021
[4]Beckman Research Institute of the City of Hope, Duarte, CA 91010

Mating pheromones are secreted into the medium by the ciliate *Euplotes raikovi*, a free living marine organism, where they act as mating type specific signals that distinguish different intraspecific classes of cells. Thus these soluble factors aid in the communication between cells that leads to conjugation by inducing formation of mating pairs following interaction with cell surface receptors.

It has been proposed (1) that ciliate mating types evolved as a mechanism of self recognition similarly to autocrine secretion systems of higher animals the cells producing the mating pheromones have functional external receptors for their own secreted molecules. This model assumes that each cell-type specific mating pheromone may bind either to receptors on cells producing the same pheromone (homologous binding) or receptors on cells unable to produce that pheromone (heterologous binding), but only the second one will be able to generate a mating reaction through a competitve mechanism leading to the loss of self recognition. Mating pheromones that have been isolated and partially characterized from ciliates are those from *Blepharisma japonicum* (2-4), *E. raikovi* (5-8), and *Euplotes octocarinatus* (9, 10). It has been shown that in these two species of *Euplotes* many different mating pheromones are produced and the polymorphism is particularly high in *E. raikovi*. They are proteins controlled by a series of alleles codominant at the Mendelian mating type (*mat*) locus (7). The analysis of the mechanism of mating type inheritance showed a diffused condition of heterozygosity in the wild-type parental strains for different codominant *mat* alleles. The allele *mat*-1 segregates with E*r*-1 (designated abbreviation to indicate *E. raikovi* mating pheromones) and confers mating type I to the cells, *mat*-2 segregates with E*r*-2 confering mating type II, etc. (11). The homozygotes produce only one E*r* type while the heterozygotes behave as a combination of the two corresponding homozygotes and produce two E*r* types as individual species (7).

Five mating pheromones (E*r*-1, E*r*-2, E*r*-3, E*r*-9, and E*r*-10) have been isolated and purified to homogeneity; the sources are *E. raikovi* cells homozygous for the allele *mat*-1, *mat*-2, *mat*-3, *mat*-9, and *mat*-10,

0097–6156/90/0434–0153$06.00/0
© 1990 American Chemical Society

respectively. The purification, initiallly developed for E$\underline{r}$-1 (6), involved a three-step procedure based on reverse-phase chromatography on Sep-pak C$_{18}$ cartridges, gel filtration on Sephadex G-50, and ion exchange chromatography on a Mono Q column. As summarized in Table 1, they constitute a homologous family of small proteins with an averaged molecular weight of 10,000, reflecting their native homodimeric structure; they are acidic, with isoelectric points in the range 3.7-4.0, and they are characterized by a variable number of amino acids with a prevalence of acidic and half-cystine residues. They all have aspartic acid as unblocked amino terminal amino acid. This is consistent with the cleavage of the mating pheromones from larger precursor molecules, as recently demonstrated for E$\underline{r}$-1 from cDNA sequence studies (12).

Table 1

Properties of <u>Euplotes</u> <u>raikovi</u> Mating Pheromones

Subunit Molecular Weight:	4,000 - 5,000
Subunit Structure:	Non-covalent homodimer
Amino Acid Content:	38 - 40 residues
	15 - 16% Cystine
	2.5 - 5% Basic residues
	12.5 - 17.5% Acidic
residues	
Isoelctric Point:	3.7 - 4.0
N-terminal:	Aspartic acid (free)
CHO:	None

The primary structure of E$\underline{r}$-1 was the first one to be completed (13) and will be described in grater detail. As shown in Fig. 1 the sequence was determined by automated Edman degradation of the whole protein and peptides generated from four different hydrolyses, cyanogen bromide, trypsin, <u>Staphylococcus</u> <u>aureus</u> V8 protein, and chymotrypsin of the carboxymethylated protein. These peptides were separated by reverse phase chromatography, subjected to amino acid analysis, and eventually automated amino sequence analysis. The analysis of the undigested carboxymethylated protein identified the first 20 amino acid residues. The cyanogen bromide cleavage yielded only one soluble peptide, constituting the carboxy-terminal portion of E$\underline{r}$-1. The insoluble fraction was further subjected to trypsin digestion and four peptides were partially sequenced. From <u>S</u>. <u>aureus</u> V8 protease digestion seven peptides were recovered and their analysis and alignment provided most of the sequence of E$\underline{r}$-1, as only one peptide was not recovered. The data to complete the sequence were provided by the amino acid sequence of a 20-residues peptide generated by a chymotrypsin digestion. The last two residues of E$\underline{r}$-1 were determined by carboxypeptidase Y digestion confirming the carboxyl-terminal sequence. The sequence determined accounts for a single chain of 40 residues.

This complex strategy, considering the relative small size of this structure, was dictated by the initial discrepancy in the expected molecular weight (12,000) and that finally reported (4,410). Determinations of the native and subunit molecular weights of E$\underline{r}$-1, as well as of the other mating pheromones, have varied over a wide range, making intrepretation difficult. A value of 12,000, was initially estimated on native samples of E$\underline{r}$-1 on a Bio-Gel P-10 gel filtration chromatography (6), but more recent data from a gel

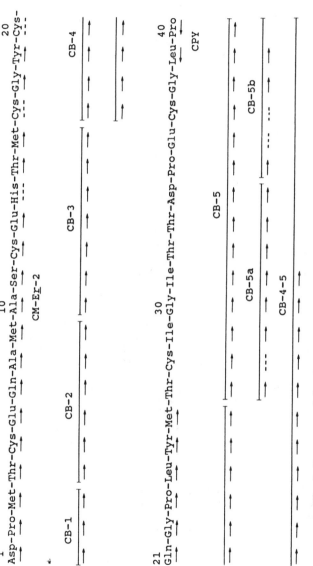

Fig. 1 Summary of the data used to establish the complete amino acid sequence of Er-1 mating pheromone. The peptides have been designated and numbered according to the type of digest and the theoretical order in which they appear in the sequence. Designations are: CNBr, cyanogen bromide; T, trypsin; V8, S. aureus V8 protease; CT, chymotrypsin. Peptides indicated by two numbers connected with a hyphen result from partial cleavage. Residues directly identified by automated Edman degradation and carboxypeptidase Y digestion (CP-Y) are marked by right and left arrows, respectively. residues identified by amino acid composition are indicated by dashed lines. "Taken from ref. 13 and reproduced by permission of the American Society of Biochemistry and Molecular Biology".

filtration performed on a Superose-12 column suggested a value of 9,000, and sodium dodecyl sulfate polyacrylamide gel electrophoresis revealed the presence of a diffused band in the range of 4,000-5,000.

A conclusive assignment of the molecular weight of the mating pheromone Er-1 was given by mass spectral analysis. A sample of native Er-1 was subjected to Cf-252 fisssion fragment ionization mass spectrometry using a mass spectrometer constructed at the Rockefeller University (14). As shown in Fig. 2 a strong protonated quasimolecular ion $(M + H)^+$ appears in the positive spectrum at m/z 4411.2, corresponding to an isotopically averaged mass of 4410.2. This value is in agreement with the calculated molecular weight of 4411.0 from the proposed sequence.

Er-1 contains 6 half-cystine residues in the native molecule and, consisting with it being an extracellular protein, they probably exist as either intra- or inter-chain disulfide bonds. The molecular weight obtained from the sequence is inclusive of the existence of three disulfide bonds and the presence of a minor peak at m/z 8819 in the mass spectrum is suggestive, considering the data from other molecular weight determinations, of a native homodimeric structure (or possibly higher aggregates) with the subunits associated by noncovalent forces.

Analysis of the other mating pheromones by mass spectrometric analysis under the same conditions used for Er-1, gave a strong protonated quasimolecular ion appearing in the positive spectrum at m/z 4297.5 for Er-2 (corresponding to an isotopically averaged mass of 4296.5) and a m/z 4190.7 for Er-10 (corresponding to an averaged mass of 4190.2) (Fig. 2). These values are in agreement with the calculated masses of 4297.1 for Er-2, and 4191.7 for Er-10, from the respective proposed sequences. The spectra of both Er-2 and Er-10 also contain, as in the case of Er-1, a secondary peak, much weaker, at m/z 8593 and 8380, respectively. This finding supports a native homodimeric structure, common to all the members of this family of proteins.

Different strategies were used for the determination of the sequences of Er-2 (Raffioni, S., Luporini, P., Miceli, C., and Bradshaw, R.A., manuscript submitted), and Er-10 (15), respectively. Automated Edman degradation of intact carboxymethylated Er-2 that permitted the identification of the first 25 residues, with only three equivocal positions, and of peptides generated by cyanogen bromide cleavage at the four methionine residues were sufficient to provide the complete amino acid sequence. Carboxypeptidase Y digestion was used to confirm the last two residues at the carboxyl terminal. The complete amino acid sequence of Er-10 (15) involved automated Edman degradation of the entire protein after performic acid oxidation which yielded the first 30 positions with the exception of six residues. Five of them were subsequently identified to be half-cystines during Edman degradation of peptides derived by cyanogen bromide and by S. aureus V8 cleavage of carboxymethylated Er-10. As only one methionine is present in this pheromone, peptides derived from cyanogen bromide cleavage were analyzed without further separation. To complete the sequence, a peptide obtained after fractionation on a reverse-phase HPLC on a C_{18} column of a digest from S. aureus of carboxymethylated Er-10 was sequenced for all of its 23 residues.

In order to assign the disulfide bonds of these molecules fast atom bombardment mass spectrometry (FABMS) which has been used not only to confirm amino acid sequence data but also to elucidate post-translational modifications of proteins, such as disulfide bonds, has been employed. For this purpose a sample of native Er-2, containing four methionines, was subjected to CNBr cleavage and without further fractionation directly

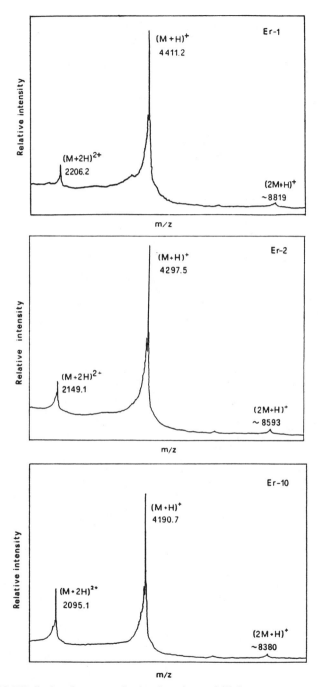

Fig. 2 Cf-252 fission fragment ionization time-of-flight mass spectra of Er-1, Er-2, and Er-10. The region between m/z 1600-10000 is shown. M designates the intact Er molecule. The data for Er-1 are taken from ref. 13 by permission of the American Society of Biochemistry and Molecular Biology.

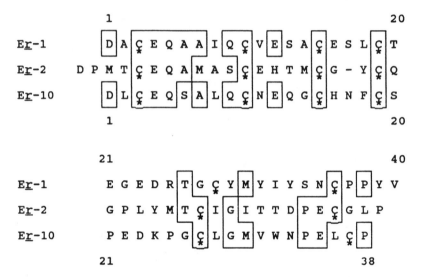

Fig. 3 Comparison of the amino acid sequences of the mating pheromones Er-1, Er-2, and Er-10. A two residue-shift and one gap have been arbitrarily introduced in Er-2 to improve the alignment with the other two pheromones. The half-cystine residues have been emphasized by asterisks.

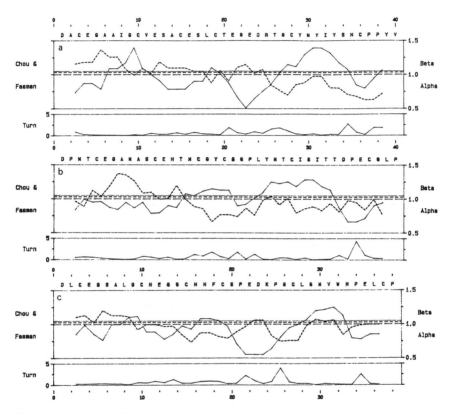

Fig. 4 Pepplot graphic of secondary structure predictions of the mating pheromones Er-1 (a), Er-2 (b), and Er-10 (c). Dotted lines are used to distinguish the curves for alpha structures from the curves for beta structures. The horizontal dotted and solid lines indicate the minimum levels for predicting beta and alpha structures. The data for Er-1 are taken from ref. 13 and Er-10 from ref. 15 by permission of the American Society of Biochemistry and Molecular Biology and the American Chemical Society, respectively.

analyzed on a JEOL HX100HF mass spectrometer. A ion of m/z of 1343.93
has been detected which is consistent with a disulfide-bonded peptide pair
composed of peptides CB-3 and CB-5 indicating a disulfide bond between
Cys II and Cys VI. Attepts to determine the other two disulfide bonds from
this cleavage have been unsuccesful. A combination of chemical and
enzymatic cleavages of the different mating pheromones and separation of
the peptide mixture by HPLC before FABMS analysis are in progress to
establish the complete disulfide pattern.

A comparison of the sequences of Er-1, Er-2, and Er-10 (Fig. 3) reveal
that the molecules are clearly homologous with the most striking similarity in
the amino terminal region. In order to align the three structures a shift of two
residues and one gap have been introduced in Er-2. They all share aspartic
acid as a commom amino terminal residue and the Cys-Glu-Gln-X-X-X-X-
Cys motif. This homologous structure is suggestive of a common function,
such the site of contact of the homodimers. On the other hand, the lack of
similarity present in the carboxyl terminal region of these molecules may
reflect a unique function such as a a receptor binding site. An analysis of the
secondary structure using the PEPPLOT programm (16) revealed a
probable helical segment in the amino terminal region of all the three
sequences and a potential reverse turn region in the carboxyl terminal end
making it a candidate for a surface exposed region (Fig. 4a-c). Even if this
analysis is probably of limited importance for such small molecules
containing three intrachain disulfide bonds, it emphasizes the similarity in
the three dimensional structure to be expected for Er-1, Er-2, and Er-10
despite the low level of identity. Such an expectation is supported by the
conservative position of the six half-cystines in all the three sequences.

Acknowledgment

This work was supported by a research grant from the U.S. PHS
DK32465 (to R. A. B.) and the Italian Ministero della Pubblica Istruzione
and RR00862 (to B. T. C.) and GM40673 (to T.L. and J.S.).

Literature Cited

1. Luporini P, and Miceli C (1986) in Molecular Biology of Ciliated Protozoa
(Gall J, ed) pp 263-299, Academic Press, Orlando, FL
2. Kubota T, Tokoroyama T, Tsukuda Y, Koama H, and Miyake A (1973)
Science 179, 400-402
3. Miyake A, and Beyer J (1974) Science 185, 621-623
4. Braun V, and Miyake A (1975) FEBS Lett 53, 131-134
5. Miceli C, Concetti A, and Luporini P (1983) Exp Cell Res 149, 593-598
6. Concetti A, Raffioni S, Miceli C, Barra D, and Luporini P (1986) J Biol
Chem 261, 10582-10586
7. Luporini P, Raffioni S, Concetti A, and Miceli C (1986) Proc Natl Acad Sci
USA 83, 2889-2893
8. Raffioni S, Miceli C, Concetti A, Barra D, and Luporini P (1987) Exp Cell
Res 172, 417-424
9. Weischer A, Freiburg M, and Heckmann K (1985) FEBS Lett 191, 176-180
10. Dieckhoff H S, Freiburg M, and Heckmann K (1987) Eur J Biochem 168,
89-94
11. Luporini P, and Miceli C (1984) Protistol 20. 371-376.
12. Miceli C, La Terza A, and Melli M (1989) Proc Natl Acad Sci USA 86,
3016-3020
13. Raffioni S, Luporini P, Chait B T, Disper S S, and Bradshaw R A (1988)
J Biol Chem 263, 18152-18159

14. Chait B T, Agosta W C, and Field F H (1981) Int J Mass Spectrom Ion Phys 39, 339-366
15. Raffioni S, Luporini P, and Bradshaw R A (1989) Biochemistry 28, 5250-5256
16. Gribskov M, Burgess R R, and Devereux J (1986) Nucleic Acid Res 14, 327-334

RECEIVED February 27, 1990

Chapter 10

Micropellicular Sorbents for Rapid Reversed-Phase Chromatography of Proteins and Peptides

Krishna Kalghatgi and Csaba Horváth

Department of Chemical Engineering, Yale University, New Haven, CT 06520

The use of micropellicular stationary phases consisting of non-porous silica microspheres with a thin alkyl-siloxane retentive layer is described for rapid HPLC analysis of proteins by reversed phase chromatography. The applications include peptide mapping, analysis of protein mixtures, assay of protein purity and the potential of short columns packed with such sorbents for process monitoring in biotechnology by reversed phase HPLC is demonstrated. The advantages of micropellicular stationary phases over conventional sorbents for rapid HPLC analysis of proteins stems from the lack of internal porosity which allows for relatively high column efficiency at high flow velocities, fast column regeneration, improved column stability at high temperatures and good sample recovery. Nevertheless, further advances in instrument design are required to exploit the full potential of columns packed with micropellicular sorbents in the HPLC of proteins by gradient elution at elevated temperature.

Chromatography has traditionally been the prime separation method in biochemistry and the introduction of rapid and more efficient HPLC techniques is expected to expand its scope to process monitoring and quality control in biotechnology. Recently, reversed phase chromatography (RPC) has become a powerful analytical technique for biological macromolecules, particularly for peptides and proteins (1). The availability of efficient columns and precision HPLC instrumentation has greatly contributed to unfold the potential of RPC for analysis of biopolymers. By virtue of its versatility, RPC has been used successfully in numerous biotechnological applications including separation and purification of peptides, determination of protein structure (1-3), measurement of purity in quality control (4-6) and process monitoring (7,8). Advances in recombinant DNA and hybridoma technologies over the past decade have led to the introduction of several proteinaceous drugs by the pharmaceutical industry (9). Isolation and purification of therapeutic proteins is demanding because of their complex structure and the processes involved in their production. The need to obtain them in very high levels of purity necessary for human use requires the removal of closely related variants and calls for efficient and high resolution techniques.

Due to recent advances in column technology, novel stationary phases have become available for such applications. This communication deals with the use of micropellicular sorbents which consist of a fluid-impervious microspherical support with a thin retentive layer at the surface. For biopolymer analysis by HPLC, such stationary

0097–6156/90/0434–0162$06.00/0
© 1990 American Chemical Society

phases offer certain advantages over the conventional porous sorbents in terms of speed of analysis, efficiency and sample recovery. The salient features of this approach and some of the problems associated with high speed HPLC are discussed and demonstrated by experiments in the analysis of proteins by reversed phase HPLC.

CHROMATOGRAPHIC SYSTEM

REVERSED PHASE CHROMATOGRAPHY. Although governed by the same chromatographic principles, RPC of proteins differs significantly from that of small molecules because of the large size and complex tertiary structure of protein molecules (10). High molecular weight is associated with low diffusivity; hence, separation of biological macromolecules under isocratic elution conditions would require very much longer time than that of small molecules. The protein molecules undergo conformational changes and/or denaturation in the mobile phase and upon contact with the non-polar chromatographic surface (11). The dynamics of the separation process may be further complicated by slow kinetics due to multipoint attachment (12), precipitation and redissolution may occur in the chromatography of proteins and other biological macromolecules (13). For proteins, RPC represents a technique distinct from the other chromatographic methods such as size exclusion, ion-exchange or hydrophobic interaction chromatography. Due to contact with strongly acidic hydroorganic mobile phases and the strongly hydrophobic chromatographic surface, the protein molecules do not traverse the column in their native form. As a result, RPC of proteins exhibits unique selectivities and relatively high efficiency of separation.

STATIONARY PHASE. In the current practice of biopolymer RPC, stationary phases prepared from rigid, macroporous, microparticulate supports with covalently bonded n-alkyl ligates with 4 to 18 carbons are employed. Silica has been the most popular support because of its availability in narrow particle and pore sizes, high mechanical stability, and reactive surface for chemical modifications (14). It is not stable at alkaline pH and the residual silanols at the surface usually have an undesirable effect on chromatography of substances with basic functional groups due to silanophilic interactions (15). Supports such as those prepared from styrene-divinylbenzene copolymers which are stable in a wide pH range, have found increasing employment in the chromatography of biological macromolecules. Since the native poymeric material is hydrophobic, it can be used without chemical modification, and columns packed with such sorbents are generally stable and can be cleaned with alkaline solutions. However, some polymeric supports swell upon contact with organic solvents resulting in poor permeability of the columns and the separation efficiency for small molecules and hence, such supports are usually inferior.

PARTICLE DIAMETER AND PORE SIZE. The columns for separation of large biological molecules are packed with 5-10 μm stationary phase particles having pore sizes in the range of 25-100 nm. The microarchitecture of macroreticular stationary phases can have significant influence on the efficiency of separation particularly for large molecules. Mesoporous supports facilitate access for the large eluite molecules to the chromatographic surface, but at present further increase in the pore size above 1000Å is accompanied by a deterioration of the mechanical properties of the supports. Furthermore, poresize distribution of many column packings is rather wide and poor recovery is frequently encountered due to entrapment of large molecules in the porous interior.

COLUMN SIZE. Analytical reversed phase columns for proteins and peptides range from 50 to 250 mm in length having 4 to 5mm i.d. Columns with 2mm i.d. have also been employed for peptide separations in microsequencing applications. The use of microbore columns with less than 2 mm i.d. is severely hampered by instrumental constraints due to small extra-column dead volumes and difficulties in obtaining accurate flow rates in the range of 10 to 100 μl/min by gradient elution.

MOBILE PHASE. In most cases, RPC of proteins is carried out with acidic hydroorganic eluents containing trifluoroacetic acid (TFA). The low pH improves efficiency by suppressing the ionization of carboxyl groups on the protein molecule as well as residual silanols on the surface. TFA is a good solubilizing agent for proteins, forms ion-pairs and is volatile. Other additives such as phosphoric, formic and perchloric acid have also been used but found limited applications in cases where they offer improved selectivity. Among the organic modifiers, acetonitrile is used most commonly because of its low viscosity and optical transparency. Propanol which has greater elution strength and less denaturing effect, is also employed frequently as the organic modifier.

GRADIENT ELUTION. The separation of proteins in RPC is carried out by gradient elution and the chromatographic results are influenced by the gradient conditions and flow rate. According to Snyder (13), a measure of retention in gradient elution is given by the equation,

$$\bar{k} = \frac{F\ t_G}{1.15\ V_m\ S\ (\Delta\ \Phi)} \tag{1}$$

where $\bar{k}$ is the retention factor of the eluite at the column midpoint, F is the flow rate, Δ F is the change in the volume fraction of organic solvent in the mobile phase during gradient elution, V_m is the column void volume, t_G is the gradient time and S is given by the slope of the linear plots of the logarithmic retention factor against F. In comparison to small molecules, proteins have larger contact area with the chromatographic surface and as a result, the retention factor is very sensitive to the concentration of the organic modifier *i.e.*, the value of S values is relatively large.

EFFECT OF TEMPERATURE. Temperature can have profound effect on the retention of proteins in RPC. Generally, retention decreases with increasing temperature with concomitant increase in column efficiency due to increased solute diffusivity. However, retention of proteins may increase under certain conditions when increase in temperature promotes further unfolding of the protein.

MICROPELLICULAR STATIONARY PHASES

Before the advent of HPLC, the chromatography of biopolymers was carried out almost exclusively with stationary phases prepared from relatively large polysaccharide based particles such as cellulose, cross-linked dextran or agarose. These stationary phases have played an enormously important role in column chromatography of biopolymers despite their limitations due to poor mechanical properties. Early attempts to improve the mechanical and mass transfer properties *e.g.* by coating celite with an ion-exchange resin layer (16,17), were met with only limited success. At the dawn of HPLC, narrow bore columns packed with pellicular stationary phases made from 40μm glass beads with a thin retentive layer on the surface were introduced (18). Due to the high mechanical strength, and relatively high efficiency of pellicular sorbents, such columns could be operated at high column inlet pressures and at relatively high flow rates without significant loss in separation efficiency. This approach was superseded by subsequent introduction of 10 μm totally porous bonded phases which had much higher efficiency and sample loading capacity, so that the further development of HPLC of small molecules proceeded by using porous microparticulate column packings.

With the recent growth of biopolymer HPLC, there has been a renewed interest in pellicular stationary phases for analytical applications. Unger and co-workers (19) described column packings made of 1.5 μm, non-porous, monodisperse silica microspheres with an alkyl siloxane layer at the surface for biopolymer separation . This was followed by the introduction of other micropellicular stationary phases based on siliceous or polymeric supports for HPLC of large biomolecules (Table I).

Table I. Micropellicular Stationary Phases for HPLC Analysis of Biopolymers

Support		Chromatographic	Ref.
Particle diameter (μm)	Type	application	
1.5	Silica	RPC	(19)
		EIC	(20)
		HIC	(21)
2	Silica	RPC	(12)
		BIC	(13)
		EIC	(24)
		HIC	(24)
		MIC	(24)
2.5	Polymeric	RPC	(26)
		EIC	(26)
3	Polymeric	RPC	(27)
		EIC	(28)
4	Polymeric	RPC	(29)
7	Polymeric	EIC	(30)

The acronyms are: RPC, reversed phase chromatography; EIC, ion-exchange chromatography; HIC, hydrophobic interaction chromatography; BIC, biospecific interaction (affinity) chromatography.

A major advantage of micropellicular sorbents stems from the rapid mass transfer for eluite exchange between the stationary and mobile phases in the column. As suggested by Figure 1, the interaction of eluites with the stationary phase ligates is highly facilitated by their confinement to a thin layer at the surface, which permits complete exposure to the mobile phase stream in the interstitial space of the column packing. In other words, the diffusional path length is very short or virtually absent in the retentive surface layer of micropellicular sorbents so that the plate height contribution of the C-term in the van Deemter equation (31), is relatively small. As a result, the fast mass transfer in the stationary phase and in the mobile phase due to the small particle size, makes it possible to obtain high column efficiency even at relatively high flow velocities.

However, columns packed with 1-2 μm particles, have certain disadvantages. Due to small particle size, the columns packed with micropellicular stationary phases have low permeability (27) and therefore, can not be operated at very high flow rates due to pressure limitations of commercial HPLC instruments. In comparison to porous particles, the surface area of stationary phases per unit column volume is low, and hence, their loading capacity is correspondingly smaller. This is particularly evident in the isocratic analysis of small molecules where the column can be easily overloaded. Therefore, micropellicular sorbents do not appear to offer advantages in the HPLC of small molecules.

On the other hand, the lack of internal pore structure with micropellicular sorbents is of distinct advantage in the analytical HPLC of biological macromolecules because undesirable steric effects can significantly reduce the efficiency of columns packed with porous sorbents and also result in poor recovery. Furthermore, the micropellicular stationary phases which have a solid, fluid-impervious core, are generally more stable at elevated temperature than conventional porous supports. At elevated column temperature the viscosity of the mobile phase decreases with concomitant increase in solute diffusivity and improvement of sorption kinetics. From these considerations, it follows that columns packed with micropellicular stationary phases offer the possibility of significant improvements in the speed and column efficiency in the analysis of proteins, peptides and other biopolymers over those obtained with conventional porous stationary phases. In this paper, we describe selected examples for the use of micropellicular reversed phase sorbents prepared from 2-μm silica microspheres in rapid HPLC analysis of proteins and peptides and discuss some of the peculiarities associated with this approach.

EXPERIMENTAL

MATERIALS. Ribonuclease A (RNase A) (bovine pancreas), cytochrome C (Cyt. C) (horse heart), lysozyme (chicken egg white), myoglobin (sperm whale), β-lactoglobulin A (β-Lact. A) (bovine milk), carbonic anhydrase (bovine erythrocytes) and trifluoroacetic acid (TFA) were purchased from Sigma (St. Louis, MO, USA). L-Asparaginase was from Merck, Sharp & Dohme (West Point, PA, USA). N-tosyl-L-phenylalanine chloromethyl-ketone (TPCK)-treated trypsin was obtained from Worthington (Freehold, NJ, USA). Human growth hormone was a gift from Genentech (South San Francisco, CA, USA), Allergen Type SQ 555 Katze was from Ephipharm (Linz, Austria). HPLC-grade acetonitrile (ACN), reagent-grade ortho-phosphoric acid, and buffer salts were from J.T. Baker (Phillipsburgh, NJ, USA). Eluents were prepared with deionized water, prepared with a NanoPure system (Barnstead, Boston, MA,USA), filtered through a 0.45-μm filter and degassed by sparging with helium before use.

COLUMNS. Hy-Tach C-18 micropellicular reversed phase columns (30x4.6mm, 105x4.6 mm), were obtained from Glycotech (Hamden, CT, USA). Experiments were also carried out by using 150x4.6 mm, or 30x4.6mm columns packed with 5-μm, Vydac C-4 silica based totally porous reversed phase material (Separations Group, Hesperia, CA, USA).

INSTRUMENTATION. A Model 1090 Series M liquid chromatograph (Hewlett Packard, Avondale, PA, USA), equipped with a ternary DR5 solvent delivery system, diode-array detector, ColorPro graphic plotter, and autosampler, was used. The chromatographic system and data evaluation were controlled by Series 79994A Chem Station computer.

APPROACHES TO FAST ANALYSIS.

Most commercially available HPLC analyzers have been designed to provide precise and accurate analytical information on the composition of samples containing small molecules by using conventional columns. The time scale of the separation by isocratic or gradient elution under these conditions corresponds to standards established about a decade ago. For this reason, the performance of these instruments is less satisfactory in applications with high speed columns which allow the analysis of biopolymers to be carried out on a time scale of a minute or less. In order to expand the scope of HPLC to such applications, some of the considerations described below should be taken in to account in the design of fast HPLC analyzers for biological macromolecules.

GENERATION OF ELUENT GRADIENT. In most instances, proteins have to be analyzed by gradient elution and high speed analysis with short column requires steep gradients. In other words, it is necessary to generate rapid and reproducible gradients on a time scale commensurate with the time of analysis. High performance mixing devices with low internal volume are preferred but it should be kept in mind that their performance may depend on other factors such as precision of eluent delivery by the pumps. In current HPLC instruments, gradients are formed either at low pressure using a single pump and electronically controlled solenoid valves, or at high pressure with two pumps by using appropriate static or dynamic mixers. Generally, mixing devices have larger internal volume than static mixers and therefore, their use is associated with greater 'gradient dead volume', *i.e.* the volume of the liquid between the point of mixing and the column entrance. The use systems with low dead volume results in short dwell time which is essential for rapid gradient formation *i.e.* fast analysis and column reequilibration.

TEMPERATURE CONTROL. Since it is widely believed that adequate and reproducible results are obtained when HPLC analysis is carried out at room temperature (36), many liquid chromatographs manufactured until recently do not have the capability to control the temperature of the column and of the eluent entering the column. If temperature control is needed, it is accomplished by the use of an oven for maintaining the column at elevated temperature or by the use of column jackets with a circulating liquid. Control of the column temperature alone is not adequate for rapid analysis by gradient elution at elevated temperature since temperature variations can occur due to heat effects associated with the mixing of eluent components (*e.g.* endothermic for ACN/water; exothermic for methanol/water), and poor heat transfer. However, in chromatography with rapid gradient elution at elevated temperature, the major problem is to heat the eluents from ambient to the temperature of the column so that no radial temperature gradients are set up inside the column. Further, it is required that all components of the system downstream from the point of mixing, such as sampling valve, guard column, eluent filter etc. are maintained at the same temperature as the column.

Although, the detectors and data handling equipment in most modern HPLC instruments are satisfactory for conventional HPLC work, for fast analysis their time constants should not exceed 100 ms. UV detectors should be designed such that changes in refractive index of the effluent do not cause excessive baseline drift when steep gradient are used. Flow cells which can withstand 1000 psi or more of back pressure are preferred with a restriction at the outlet to prevent bubble formation when the analysis is carried out at elevated temperature. The above guidelines have been followed in the construction of the instrument from components described in the experimental section and illustrated in Figure 2. The gradient delay volume of the instrument was approximately 0.6 ml, which was adequate at the flow rate used in the various applications presented here. The dead volume is still too large for use of narrow bore columns having inner diameter 1 mm or less.

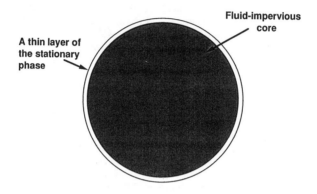

Figure 1. Schematic representation of the micropellicular stationary phase.

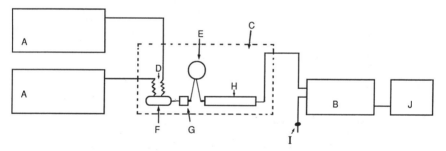

Figure 2. Flow Sheet of the HPLC Unit for Rapid Protein Analysis. A binary gradient system with gradient controller, Model 21500 pumps,Model 2152 controller, Pharmacia, Piscataway, NJ, U.S.A. (A), Variable wavelength UV-Visible detector, Model LC 95, Perkin Elmer, Norwalk, CT, U.S.A (B), Constant temperature circulating bath, Model DL-8 Haake Buchler, Saddlebrook, NJ, USA (C), Heat exchangers (D), Sampling valve, Model 7125 injector, Rheodyne, Cotati, CA, U.S.A (E), Gradient mixer (10µl internal volume), Lee Company, Westport, CT, U.S.A. (F), Pulse dampner (G), Column (H), Flow-cell restrictor (I), C- R3A Chromatopak integrator, Shimadzu, Columbia, MD, U.S.A.(J).

The chromatograms presented in Figures 4 and 8-12 were obtained by using the HPLC unit depicted in Figure 2   Results shown in the other figures were from experiments carried out with Hewlett Packard Model 1090 liquid chromatograph.

RESULTS AND DISCUSSION

RAPID ANALYSIS OF PEPTIDES AND PROTEINS.   Columns packed with micropellicular stationary phases can be utilized in many areas of protein chemistry and biotechnology where high analytical speed and resolution are required.   Peptide mapping by HPLC is routinely employed for the determination of structure of proteins (2, 3, 32). In the manufacture of proteinaceous pharmaceuticals, peptide mapping by RPC is a widely employed tool for measurement of protein purity and also as a method of quality control.   Such industrial applications usually involve large number of samples and their handling by conventional HPLC methods requires considerable time.   A significant reduction in the time of analysis can be achieved by using columns packed with micropellicular sorbents which facilitate peptide mapping in relatively short time as shown by the chromatograms of tryptic digests of carbonic anhydrase, L-asparaginase, and myoglobin, in Figure 3.   In each case, the chromatographic runs including column regeneration were completed in about half an hour, which is considerably faster than the the time required by using columns packed with porous stationary phases.

ASSAY OF PURITY.   Despite relatively low sample loading capacity, short columns packed with micropellicular reversed phase sorbents can be used for the rapid measurement of trace impurities in protein samples.   When sufficiently large sample loads and gradient elution are used, the minor components are separated and detected with adequate sensitivity.   This is illustrated by the chromatograms of three proteins in Figure 4 and the chromatographic runs were completed in a few minutes.   From the measurement of the peak area, in the cases illustrated the purity of carbonic anhydrase, L-asparaginase, and myoglobin was found to be 90.0, 87.6 and 78.3%, respectively.

COLUMN STABILITY.   The absence of a porous support structure results in enhanced column stability at elevated temperature and pH even with micropellicular sorbents prepared from siliceous supports (14).   This is illustrated by the chromatogram in Figure 5 which shows the separation of minor conformers of human growth hormone by using a moderately alkaline mobile phase (pH 8.5).   Prior to obtaining the above chromatogram, the column was perfused with 4000 column volumes of the mobile phase at 80°C, yet no noticeable changes in retention behavior, separation efficiency and sample recovery had been observed with respect to initial column performance.

PROCESS MONITORING.   Rapid analytical HPLC appears to have a great potential in the monitoring of bioprocesses in which the protein or peptide composition in the process stream or reaction mixture undergoes changes.   Examples are: monitoring the column effluents in preparative chromatography of proteins or peptides, secretion of proteins in cell culture/fermentation processes, or time course of reactions in which a certain protein is modified by enzymatic or chemical methods.   In the case described here, we have monitored changes during oxidation of RNase A by performic acid, a powerful oxidizing agent widely used for quantitative analysis of cysteine and methionine residues in proteins.   However, the oxidation of proteins by performic acid is not very selective and in addition to the sulfur containing amino acids, tryptophan, tyrosine, serine and threonine are also attacked.   Although the RNase A molecule lacks tryptophan, it contains a relatively large number of other residues which are susceptible to degradation by performic acid and yields a complex mixture of oxidized derivatives.   The reaction was carried out in a solution containing RNase A, (100 µl), performic acid (100 µl), formic

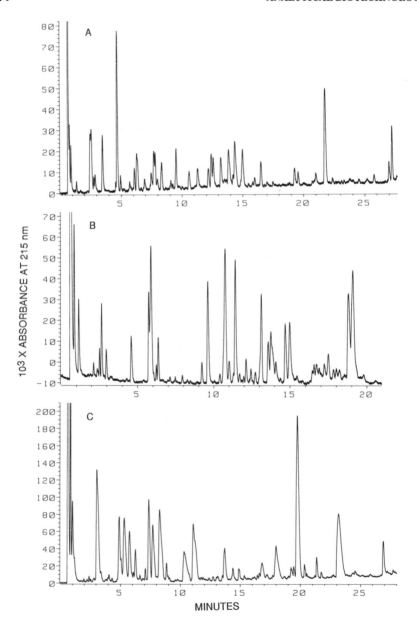

Figure 3. Tryptic maps of carbonic anhydrase (A), L-asparaginase (B) and myoglobin (C). Column: Hy-Tach micropellicular C-18 silica, 105x4.6mm; eluent A, 20 mM phosphoric acid adjusted to pH 2.8 with NaOH, eluent B, 60% (v/v) ACN, 20 mM phosphoric acid, pH 2.8; flow rate, 1.0 ml/min.; temp., 50°C. Initial column inlet pressure, 278 bars. Protein samples were carboxymethylated and subsequently digested with trypsin following the procedure of Stone *et. al.* (37). Sample, 20 μg of trypsin digest in 10 μl. Elution conditions were, 0 to 50%B (A and C) and 0 to 70%B (B) in 25 min.

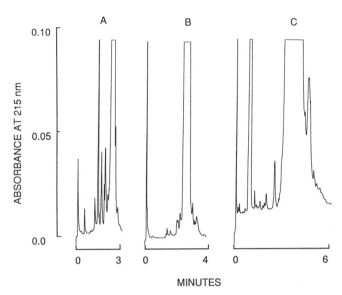

Figure 4. Determination of protein purity. Column: Hy-Tach micropellicular C-18 silica, 30x4.6mm; eluent A, 0.1% (v/v) TFA in water, eluent B, 95% (v/v) ACN in water containing 0.1% (v/v) TFA ; flow rate, 2.0 ml/min.; temp., 25°C.; Initial column inlet pressure, 270 bars. Sample, 20 μl containing 40 μg of carbonic anhydrase (A), L-asparaginase (B) and myoglobin (C). Elution conditions were, 15 to 55% B in 3 min. (A), 30 to 40% B in 4 min. (B), and 23 to 45% B in 6 min.(C).

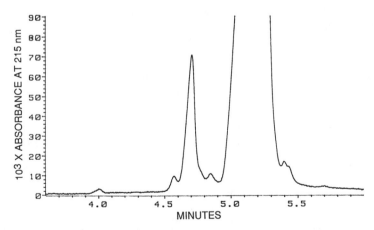

Figure 5. Reversed phase chromatography at alkaline pH. Column: Hy-Tach micropellicular C-18 silica, 30x4.6mm; eluent A, 50 mM ammonium bicarbonate in water, pH 8.2, eluent B, 60% (v/v) ACN in water containing 50 mM ammonium bicarbonate; flow rate, 2.5 ml/min.; temp., 80°C. Initial column inlet pressure, 237 bars. Sample, 25 μg of human growth hormone was preheated at 80°C for 30 min. Gradient was 65 to 80% B in 3 min.

acid (50 μl), and methanol (10 μl) at 5°C following to the procedure of Hirs (33).
Aliquots of the reaction mixture were removed at various time intervals, mixed with equal
volume of cold water and 20μl samples were then analyzed by HPLC under conditions
described in Figure 6. As seen in the chromatogram, the native enzyme elutes as a single
peak at 2.6 min. The chromatogram of the reaction mixture shows that oxidation of
methionine and/or cysteine residues begins almost instantaneously upon addition of
formic acid and the product which appears after one minute of reaction time has a shorter
retention (2.4 min) than the intact RNase A. (Figure 6B). This peak is believed to
represent the primary reaction product(s) containing methionine sufoxide and/or cysteic
acid residues. As the reaction progresses further, the concentration of the native enzyme
decreases with concomitant increase in the number of oxidation products. It is seen in
Figure 6 that all the components of the reaction mixture were eluted between 2 and 3
minutes. No attempts were made to characterize various forms of the oxidized enzyme
and the results are presented here only to exemplify the potential of rapid HPLC in such
applications.

ALLERGENS. Allergens are antigenic substances derived from diverse sources such as
pollens, mites, fungi, insects, food, drugs etc. and induce an immediate hypersensitivity
upon contact in sensitive individuals (34,35). Most allergens are considered to be
proteins, glycoproteins or pure polysaccharides in a size range of 5,000 to 70, 000
daltons. Crude extracts of allergens are employed for testing the sensitivity of allergic
patients and for hypersensitization. Current work in our laboratory is directed to the
development of chromatographic methods for purification and characterization of such
compounds. Such methods are needed for the formulation and standardization of
allergen extracts and for the isolation of pure allergenic substances to be used in the study
of structure/activity relationships. A chromatogram illustrating the separation of various
components in a commercial sample of cat allergens is shown in Figure 7. The
separation was completed in less than 15 minutes with satisfactory resolution. Further
isolation and characterization of biologically active components present in such samples is
in progress.

FAST PROTEIN ANALYSIS. Previous work from our laboratory (22-24, 27, 28) has
shown that significant reduction in the separation time for separation of proteins and
peptides can be achieved by using short columns packed with non-polar micropellicular
sorbents at high flow rates and at elevated temperature. The actual time for biopolymer
analysis, however, is the sum of the times required for the separation and for
regeneration of the column after completion of the gradient run. Therefore, for rapid
analysis, it is necessary to reduce the time required for column reequilibration. An
example of rapid protein analysis is shown in Figure 8. The analysis was carried out by
using the HPLC unit depicted in Figure 2, at high flow rate and with a steep gradient of
ACN at 80°C. The gradient profile i.e. concentration of ACN as a function of time
during the chromatographic run, was determined by a tracer technique described
previously (23). As seen in Figure 8, the analysis time for separation of proteins was
about 36 sec. and the post gradient column equilibration took about 12 sec , so that the
time required for complete analytical cycle was 48 seconds. Since the total analysis time
is proportional to the system delay time, it can be further reduced by decreasing the
system delay volume. The sample was injected manually at one minute intervals during
which both separation and column regeneration were completed as illustrated in Figure 9.
The reproducibility of results by observation of the chromatograms does not appear to be
high due to manual injection; yet, the results of 13 consecutive injections of the sample
are summarized in Table II, are satisfactory.

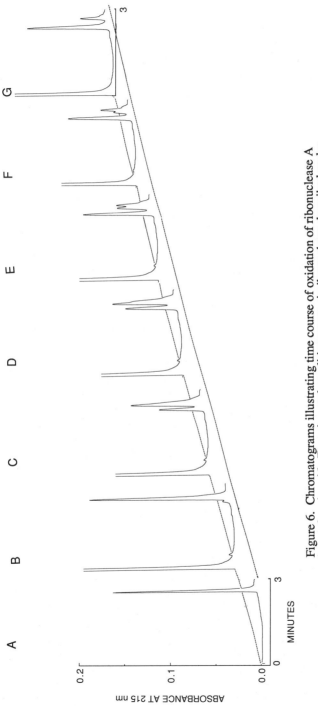

Figure 6. Chromatograms illustrating time course of oxidation of ribonuclease A by performic acid. Experimental conditions were similar to those described under Figure 3. The reaction was monitored by RPC of the aliquots (≈ 7 μg) removed at 0 (A), 1 (B), 7 (C), 12 (D), 24 (E), 48 (E), 60 (F) and 720 (G) min. Flow rate, 2.8 ml/min., temp.,30 °C. Gradient was 15 to 30% B in 3 min.

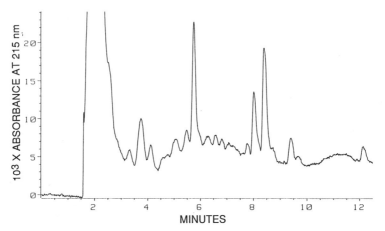

Figure 7. Protein profiling of an allergen extract. Column: Hy-Tach
micropellicular C-18 silica, 105x4.6mm; eluent A, 20 mM phosphoric acid adjusted
to pH 2.8 with NaOH, eluent B, 60% (v/v) ACN, 20 mM phosphoric acid, pH 2.8;
flow rate, 0.4 ml/min.; temp., 25°C, sample, 25 µl of allergen #SQ 555 Katze.
Gradient was 0 to 20% B in 15 min.

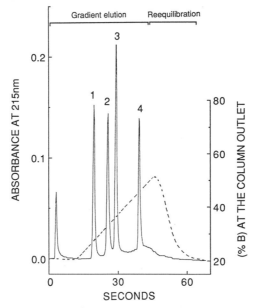

Figure 8. Fast analysis of proteins. Column: Hy-Tach micropellicular C-18 silica,
30x4.6mm; eluent A, 0.1% (v/v) TFA in water, eluent B, 95% (v/v) ACN in water
containing 0.1% (v/v) TFA ; flow rate, 4.0 ml/min.; temp., 80°C.; Initial column
inlet pressure, 260 bars; sample, 15 µl of a mixture containing ≈1 µg each of
ribonuclease A (1), cytochrome C (2), lysozyme (3) and β-lactoglobulin A (4).
Elution was carried out with a gradient 20 to 50% B in 0.5 min and return to
starting conditions in 0.1 min. The dotted line represents gradient profile of ACN.
The analysis was carried out with the house built protein analyzer described under
Figure 7.

Table II. Repetitive Analytical Cycles for Fast Protein HPLC

Protein	t_r (sec)	S.D.	R S.D. (%)	Peak Height (μV)	S.D.	R S.D (%)
RNase A	16.8	1.0×10^{-2}	3.6	10938	630	5.7
Cyt. C	22.2	6.0×10^{-3}	1.6	10396	750	7.2
Lysozyme	25.2	5.6×10^{-3}	1.3	15275	1059	6.9
β-Lact. A	34.8	4.4×10^{-3}	0.7	9446	496	5.2

The excellent reproducibility of the retention times of the late eluting sample components suggest that the performance of the system in rapid analysis is acceptable. Yet, the overall results are believed to be less reproducible than those could be obtained with such columns with an appropriately designed equipment and automated sample injector, which had yielded higher reproducibility of peak height, an essential criterion in quantitative analysis.

MICROPELLICULAR AND POROUS STATIONARY PHASES. In order to compare the features of micropellicular and porous stationary phases in rapid protein HPLC, experiments were conducted with two columns of similar size, each one of which was packed with different stationary phase and operated under comparable conditions. In this experiment the results of which are shown in Figure 10, the operational conditions were optimized for the micropellicular stationary phase (conditions A) and used subsequently for separation of the same mixture with the porous stationary phase under identical conditions. Thereafter, the elution conditions were optimized for the column packed with the porous stationary phase (condition B) and the experiment was repeated with the column packed with micropellicular stationary phase. The chromatograms are depicted in Figures 10 and 11 and the results of the two approaches are summarized in Table III. The performance of micropellicular stationary phase in either case is superior when compared by the peak width of sample components or the resolution of the critical eluite pair (Cyt.C / Lys) in the sample.

As mentioned before, the columns packed with micropellicular stationary phases have negligible intraparticulate void volume and after the gradient run, can be reequilibrated to starting conditions much faster than columns of comparable size packed with porous supports. Since the chromatographic retention of proteins and other large molecules is highly sensitive to small changes in the concentration of gradient former (13), the retention time of an early eluting component can be used to monitor the equilibration of the column. This approach was examined by measurement of retention time after injection of RNase A at various time intervals during reequilibration of the column after a gradient run. The extent of column reequilibration was evaluated by the regeneration factor defined as

$$\text{Regeneration Factor} = (t_r - t_0) / (t_r^* - t_0) \qquad (2)$$

where t_0 is column void volume, t_r & t_r^*, are the retention times of RNase measured during and after complete equilibration of the column, respectively. Plots of column regeneration factor as a function of column regenerant volume after adjustment for the

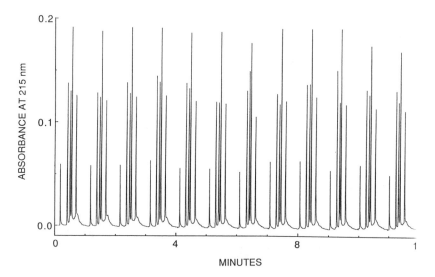

Figure 9. Rapid repetitive analysis of proteins. Experimental conditions were same as described under Figure 8 except that sample injections were made manually at 60 second intervals.

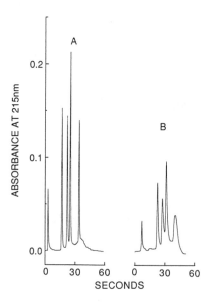

Figure 10. Comparison of micropellicular and porous stationary phases for rapid separation of proteins. Columns: Micropellicular, Hy-Tach C-18 silica, 30x4.6mm, (A) and porous, Vydac, 5 μm C-4 silica,30x4.6mm, (B). Initial column inlet pressure for the porous column was 105 bars. Experimental conditions were same as described under Figure 8

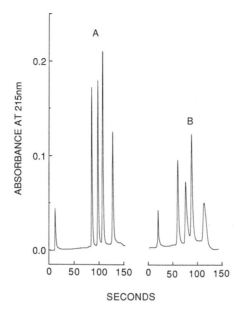

Figure 11. Comparison of micropellicular and porous stationary phases for rapid separation of proteins. Experimental conditions were same as described under Figure 9 except that the flow rate was 2 ml/min. and the gradient conditions were, 20 to 50% B in 1.5 min., 50 to 60% B in 2 min. Initial column inlet pressure was 132 and 55 bars, for micropellicular and porous columns, respectively.

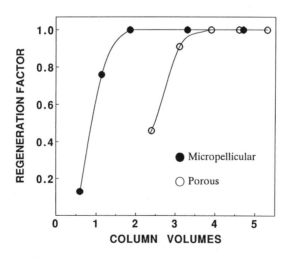

Figure 12. Reequilibration of columns with micropellicular or porous stationary phases during gradient elution. Columns: Micropellicular, Hy-Tach C-18 silica, 135x4.6mm; porous, Vydac, C-4 silica, 135x4.6mm ; flow rate, 1.0 ml/min.; temp., 80oC. Eluents were same as described in Figure 10. The columns were first equilibrated with 90% B and reequilibration was begun by a sudden drop in the concentration of B eluent to 22% in 0.1 min. Retention time of ribonuclease A was measured by gradient elution (22 to 90% B in 5 min) during column regeneration.

Table III. Performance Characteristics of Silica based Micropellicular and Wide-Pore Stationary Phases for in Rapid Separation of Proteins by Reversed Phase Chromatography

Stationary Phase	Elution Conditions[a]	Retention Time and Peak Width[b] (sec)								Resolution[c] (Cyt.C-Lys)
		RNase A		Cyt. C		Lysozyme.		β-Lact. A		
2-μm, micro-pellicular C-18	A	16.8	0.9	22.2	0.9	25.2	0.9	34.8	0.9	1.76
5-μm, 300Å porous, C-4	A	15.0	3.8	22.8	1.9	27.6	2.2	31.2	2.4	0.92
2-μm, micro-pellicular C-18	B	82.8	2.1	94.8	2.2	104.4	2.2	124.8	3.3	2.75
5-μm, 300Å porous, C-4	B	61.2	5.6	75.6	5.3	87.0	4.9	111.0	9.8	1.68

a) descibed in Figures 8 (A) and 11 (B)

b) measured at half height, $W_{0.5}$

c) calculated by using the equation $R_s = 1.175 \left\{ \dfrac{t_r(lys) - t_r(cyt)}{W_{0.5}(lys) + W_{0.5}(cyt)} \right\}$

system dead volume, are shown in Figure 12. Clearly, the column packed with the porous stationary phase required almost twice the amount of regenerant to return to the column starting conditions than the column packed with the micropellicular phase sorbent.

ACKNOWLEDGMENTS

The authors wish to thank W. S. Hancock and J. Varga for the gift of human growth hormone and the allergen sample, respectively. Thanks are also due to D. Corradini and T. Ogawa for help in some of the experiments. This work was supported by the National Foundation for Cancer Research and by Grant GM 20993 from the National Institute of Health, US Department of Health and Human Services.

LITERATURE CITED

1.   Frenz, J., Hancock, W., Henzel, W. and Horváth, Cs. HPLC of Biological Macromolecules, Methods and Applications; Gooding, K.M., Regnier, F.E. Eds.; Marcel Dekker: New York, 1990, p 145.
2.   Methods in Protein Sequence Analysis; Whittman-Liebold, B. Ed.; Springer Verlag,New York, 1989, p 220.
3.   Techniques in Protein Chemistry; Hugli; T.E. Ed.; Academic Press, San Diego, 1989, p 339.
4.   Hancock, W.S. Chromatogr. Forum 1986, 2, 57.
5.   Borman, S. Anal. Chem. 1987, 5, 969A.
6   Garnick, R.L., Solli, N.J.and Papa, P.A. Anal. Chem. 1988, 60, 2546.
7.   De Phillips, P.A.; Yamazaki, S.; Leu, F.S.; Buckland, B.C.; Gbevonyu,K.; Sitrin, R.D. Proceedings of the 9th International Meeting on HPLC of Proteins, Polypeptides, & Polynucleotides;, Philadelphia, Nov. 1989, paper, #501.
8.   Eastman, D.; Lucchesi, J.; Larsen, B.; Mullkerrin, M.; Jain, M.and Anicetti, V. Proceedings of the 9th International Meeting on HPLC of Proteins, Polypeptides, & Polynucleotides ; Philadelphia, Nov. 1989, paper, #816.
9.   Builder, S.E.,and Hancock, W.S. Chem.Eng. Progress, Aug.1988, 42.
10.   Melander, W., and Horváth, Cs. High Performance Liquid Chromatography- Advances and Perspectives; Horváth, Cs. Ed.; Academic Press, San Diego, 1980, Vol. 2, p 114.
11.   Cohen, S.A, Benedek, K., Tapuhi, Y., Ford, J.C., and Karger, B.L. Anal.Biochem. 1985, 144, 275.
12.   Velayudan, A., and Horváth, Cs. J. Chromatogr. ,1988, 443, 13.
13.   Snyder, L.R., and Stadalius, M.A. High Performance Liquid Chromatography- Advances and Perspectives; Horváth, Cs. Ed.; 1986, Academic Press, Orlando, Vol. 4, p 195.
14.   Unger, K.K., Anspach, B., Janzen, R., Jilge, G., and Lord, K.D. High Performance Liquid Chromatography- Advances and Perspectives; Horváth, Cs. Ed.; Academic Press, San Diego, 1988, Vol. 5, p-2.
15.   Kohler, J., and Kirkland, J.J. J. Chromatogr., 1987, 125, 385.
16.   Feitlson, J. and Partridge, S.M. Biochem. J., 1956 ,64, 607.
17.   Boardman, N.K. J. Chromatogr. ,1959, 2, 398.
18   Horváth, Cs. Methods of Biochemical Analysis; Glick, D , Ed.; Wiley, New York, 1973, Vol. 21, p. 79.
19   Janzen, R., Unger, K.K, Jilge, G. Kinkel J. N.and Hearn, M.T. W. J. Chromatogr.,1986, 359, 61.
20   Jilge, G.,Janzen, R., and Unger, K.K. Proceedings of the 7th International Symposium on HPLC of Proteins, Peptides and Polynucleotides; 1987, Washington, D.C.

21   Unger, K.K, Giesche, H., Kinkel, J. N and Hearn, M.T. W. J. Chromatogr. 1987, 397, 91.
22   Kalghatgi, K. and Horváth, Cs. J. Chromatogr. 1987, 398, 335.
23   Varady, L., Kalghatgi, K. and Horváth, Cs. J. Chromatogr.1988, 458, 207.
24   Kalghatgi, K. J. Chromatogr. 499, 1990, 267.
25   Kato, Y., Kitamura, T., Mitsui, A., and Hashimoto, T. J. Chromatogr, 1987, 398 ,327.
26   Kato, Y., Kitamura, T., Mitsui, A., Yamasaki, Y, and Hashimoto, T. ; J. Chromatogr. 1988, 447, 212.
27   Maa, Y. F. and Horváth, Cs. J. Chromatogr. 1987, 445, 71.
28   Maa, Y. F. Lin, S.C., Horváth, Cs., Yang, U. C, and Crothers, D.M. J.Chromatogr. 1990, 508, 61.
29   Lee, D.P. J. Chromatogr. 1987, 443, 143.
30   Burke, D.J., Duncan, J.K., Dunn, L.C., Cummings, L., Siebert, C.J. and Ott, G.S. J. Chromatogr. 1986, 353, 425.
31. Antia, F. and Horváth, Cs., J. Chromatogr. 1988, 435, 1.
32. Regnier, F. LC.GC 1988, 5, 393.
33. Hirs, C.H.W. J. Biol. Chem. 1956, 219, 611.
34. Varga, J.M., and Ceska, M. J. Allergy Clin. Immunol. 1972, 49, 274.
35   Baldo, B. A. Allergy,1983, 38, 535.
36. Parris, N.A. Instrumental Liquid Chromatography, J. Chromatogr. Library; Elsevier Scientific, New York, 1976, vol.5, p 58.
37. Stone, K.L., Lopresti, M.B., Crawford, J.M., DeAngelis, R, and Williams, K.R. Practical Guide to Protein and Peptide Purification for Microsequencing. Matsudaira, P. Ed.; Academic Press, 1989, 31.

RECEIVED April 30, 1990

Chapter 11

Displacement Chromatographic Separations on β-Cyclodextrin–Silica Columns

Gyula Vigh, Gilberto Quintero, and Gyula Farkas

Chemistry Department, Texas A&M University, College Station, TX 77843

The feasibility of the displacement chromatographic separation of positional isomers, cis/trans isomers and enantiomers is demonstrated here. In this preparative separation method the unique selectivities of the hydrolytically stable cyclodextrin–silica stationary phases and the self-focusing operation mode of displacement chromatography are combined. In order to develop a displacement chroma ographic separation, the retention behaviour of the selected solutes and prospective displacers has to be studied in the elution mode, followed by the determination of their adsorption isotherms and the realization of the actual displacement chromatographic separations. Successful separations have been achieved on alpha- and beta-cyclodextrin–silica columns, in both reversed-phase and normal-phase modes. Success of the displacement chromatographic separations critically depends on the availability of appropriate displacers and the knowledge of their adsorption isotherms. It is expected that displacement chromatographic separations on cyclodextrin–silicas will become indispensable laboratory (and hopefully, industrial) separation methods of modern pharmaceutical and biomedical research.

The potencies, pharmacological actions, plasma-disposition kinetics, metabolic fates and metabolic rates of drug enantiomers are often very different. For example, in the US, one of the most frequently prescribed beta-andrenergic blocking agents is propranolol. (+)-D-Propranolol is hundred-times as potent as the other enantiomer and its half-life in human plasma is much longer (1-3). Similar differences exist in the biological activities of the enantiomers of other cardiac beta-blockers as well (e.g. acebutolol and diacetolol) (4). The (S)-(-)-enantiomer of warfarin, an effective anti-coagulant, is three-times as active as the (R)-(+)-isomer, but

0097–6156/90/0434–0181$06.00/0

its half-life in plasma is much shorter (5). The two enantiomers have entirely different metabolic pathways and products (6). l-Fenfluramine is an extremely potent anorectic agent, while d-fenfluramine has no anorectic activity at all (7). The list of examples could go on (8).

Often, the resolution of the enantiomers represents the most difficult and costly step in the production of a new pharmaceutical. Since, on the average, only one molecule out of every ten thousands tested becomes a useful pharmaceutical, it is imperative that efficient, readily applicable separation methods be available as alternatives to the enantiospecific synthetic efforts to produce sufficient quantities of the prospective drug enantiomers in the earliest stages of drug development. Chiral separations by chromatography can be realized via diastereomer formation, the use of chiral mobile-phase additives, or the use of chiral stationary phases (9). For preparative purposes the latter ones are more practical. The most widely used and researched chiral stationary phases in liquid chromatography include the Pirkle-type phases (10), the cyclodextrin-silicas (11,12), and the derivatized cellulose-based stationary phases (13). Often, the chiral solute has to be derivatized for Pirkle-type separations, a disadvantage in preparative chromatography. Cellulose-based stationary phases are frequently hindered by solvent compatibility constraints. Cyclodextrin silicas, on the other hand, do not require derivatization of the solutes and are less constrained by mobile phase limitations. Therefore, cyclodextrin-based stationary phases are invaluable when the purpose of the separation is to produce pure enantiomers for drug developmental research and testing.

Cyclodextrin-Silica Stationary Phases. Since excellent reviews deal with the preparation, properties, and analytical applications of cyclodextrin-silica stationary phases (11,12,14-16), the following paragraphs will discuss these topics very briefly, only to the extent that the information will be used in the rest of this chapter.

Cyclodextrins are toroidally shaped molecules that contain 6,7 or 8 glucose units (alpha-, beta- and gamma-cyclodextrins). They have a hydrophobic interior cavity and a hydrophilic exterior due to the presence of 2-and 3-hydroxy groups at the larger lip and 1-hydroxy groups at the smaller lip of the cavity (11,12). Cyclodextrins can form 1:1 and 1:2 guest-host complexes (17) with molecules which penetrate their cavities. Complex stability depends on the "snugness" of the fit and the strength of the intermolecular interactions between guest and host molecules (18). The hydroxyl groups can be derivatized to yield cyclodextrins with different cavity sizes and modified intermolecular interactions.

The early polymeric cyclodextrin stationary phases had high loading capacities but lacked chromatographic efficiency (19). They were soon followed by silica-based HPLC-grade materials: by the low-capacity, hydrolysis-susceptible amino-bonded cyclodextrins (20,21) and the more stable, higher capacity alkylbonded cyclodextrin-silicas of Armstrong (22-24). The latter phases are now commercially available from ASTEC (Whippany, N.J.) under the trade-name Cyclobond (25).

The chromatographic separation of positional isomers (26-31), geometrical isomers (27,32-36) and enantiomers (37-49) has been achieved by utilizing the concerted action of inclusion complex formation, additional primary and secondary hydrogen-bond formation and steric hindrance effects between the solutes and the cyclodextrins (11,12,14-23,50). There is an abundant literature on the analytical applications of cyclodextrin-silicas (13-50), but not on their preparative chromatographic use.

Cyclodextrin-silicas have three major liabilities in preparative chromatographic separations in the elution mode: (i) the chiral selectivity factors are generally low, (ii) the loading

capacity of the phase is much smaller than that of the other
silica-based stationary phases and (iii) nonselective solute
retention is often strong. Therefore, the usual preparative
separation strategy, which requires large selectivity factors and
the use of gradient elution, cannot be applied. However, some of
these problems can be mitigated, as first introduced by Vigh et al.
(67-69), when cyclodextrin silicas are used in the displacement
mode.

Displacement Chromatography. Though displacement chromatography has
been known for many years (51), it was revived only recently, when
Horvath et al. demonstrated that efficient separations could be
achieved using microparticulate stationary phases and modern HPLC
equipment (52-58). Since then, there has been an ever increasing
interest in its utilization and several research groups pursue its
theoretical and practical aspects (59-66).
    There are three main operation steps involved in a
displacement chromatographic separation. First, the column is
equilibrated with the carrier solution. The carrier solution has
a low affinity for the column and serves as a solvent for both the
sample and the displacer. The composition of the carrier solution
is such, that the solutes, introduced in the next step, become
strongly retained at the top of the column, and are not eluted
(preferably) at all. Sample application is followed by the
introduction of the displacer, generally dissolved in the carrier
solvent. The displacer has the strongest affinity for the
stationary phase. As the displacer is continuously fed into the
column, its front begins to move down the column pushing the sample
components which, in turn, push each other according to their
adsorption strength. Presuming that the sample components are
adsorbed differently and the separation efficiency of the column
is sufficiently large, the solutes will eventually occupy adjacent
zones which move with the same velocity in the fully developed
displacement train. The equilibrium concentration of each component
in the fully developed train depends only on their respective
adsorption isotherms and the concentration of the displacer. The
exit concentration of the solutes can be much higher than what is
customary in the elution mode, a definite advantage in preparative
separations.
    While much has been learned about the role and selection of
the operation parameters in displacement chromatography (60-66),
little is known yet about the rules of displacer selection and the
means available to control the selectivity of the separation. The
paucity of well characterized displacers and the lack of knowledge
of the solute adsorption isotherms hinder most strongly the wider
acceptance and use of displacement chromatography. In most cases,
displacer selection is still done by trial-and-error. In the
majority of modern displacement chromatographic publications a
reversed-phase system was used to separate small polar molecules,
antibiotics, oligopeptides and small proteins (52-66).
    Recently, we succeeded in combining the unique separation
selectivity of cyclodextrin-silicas and the preparative efficiency
of displacement chromatography (67-69). The early results suggest
that displacement chromatography on cyclodextrin silicas may play
an important role in the development of new (mostly chiral) drugs.
In this chapter we will briefly describe the steps that we follow
in developing such separations (study of the retention controlling
factors and the adsorption characteristics of the displacers),
followed by the discussion of a few representative displacement
chromatographic separations.

Experimental

A computer-controlled displacement chromatograph was built from
commercially available components (Fig. 1) to perform the elution-

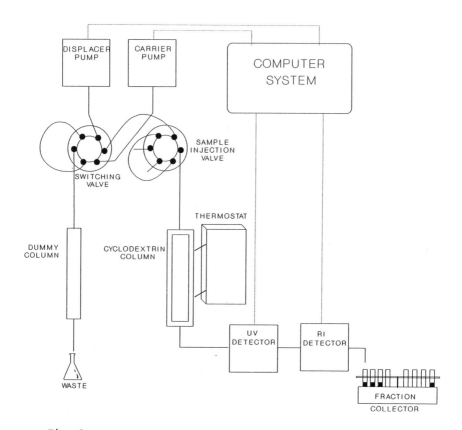

Fig. 1
Schematic of the multipurpose liquid chromatographic
system used in the present studies. For details see text.

mode, frontal-mode and displacement-mode experiments (56,59,67-69).
Two LC-2010 reciprocating piston-type pumps (Varian, Walnut Creek,
CA), capable of delivering both low and high flow rates, were
connected to the columns via a pneumatically operated, computer
controlled Type 7010 multiport switching valve and a Type 7000
injection valve (both from Rheodyne, Cotati, CA). The purpose of
the dummy column was to insure that all solutions used in the
breakthrough measurements (70) and the displacement chromatographic
separations were precompressed to the operational pressure when
column-switching occurred. 4.6 mm ID, 1/4 in OD ss. columns were
used in all experiments. Column length in the isotherm
determinations was varied according to the adsorption strength of
the solutes and the displacers. All columns were thermostatted at
30 C by water-jackets. The columns were connected to an RI-3
differential refractive index detector and an LC-2100 variable
wavelength UV detector (both from Varian). A Chrom II A/D board
(Metrabyte, Taunton, MA), connected to an IBM AT-compatible
Powermate II personal computer (NEC, Computer Access, College
Station, TX) was used for data acquisition. An interactive program,
Chromplot1, developed in our laboratory and written in Quickbasic
(Microsoft, Redmont, WA) was used to collect the data and analyze
both the chromatograms and the breakthrough curves. The nonlinear
regression routines of the SAS PC program package (SAS Institute,
Cary, NC) were used to determine the numeric parameters of the
adsorption isotherms. During the displacement chromatographic
separations fractions were collected by a Cygnet fraction collector
(Isco, Lincoln, NB). The fractions were analyzed by an LC 5560
liquid chromatograph, equipped with a Type 8085 autosampler and a
Vista 402 data-station (Varian).
        In order to eliminate the effects of occasional batch-to-
batch variations in the packing material, a single large batch of
the commercially available 5 um cyclodextrin silica, Cyclobond I,
Batch Number 880720A (ASTEC, Whippany, NJ) was used throughout the
studies. The columns were slurry packed with a Haskel pump in our
laboratory.
        The solutes and displacers (Aldrich, Milwaukee, WI and Sigma,
St. Louis, MO) were used as received, without further purification.
HPLC-grade methanol and acetonitrile (Mallinckrodt, Paris, KY), as
well as water produced by a Millipore Q unit (Millipore, Bedford,
MA) were used to make all solutions.

Results

The development of a displacement chromatographic separation
generally involves the following main steps:
   (i) determination of the retention behavior of the solutes (in
       infinite dilution) in the elution mode in order to find
       the composition of the carrier solution that provides
       sufficient initial solute retention (preferably k'>10);
  (ii) search for displacer candidates by considering the structure
       and the probable chromatographic interactions of the
       solutes;
 (iii) determination of the retention behavior (in infinite
       dilution) of the displacer candidates and elimination of
       those which are less retained than the most strongly
       retained solute of the sample;
  (iv) determination (whenever possible) of the individual excess
       adsorption isotherm of the most retained solute (by, e.g.
       frontal chromatography);
   (v) determination of the adsorption isotherms of the displacer
       candidates selected in the previous retention tests;
  (vi) selection of a displacer whose isotherm does not intercept
       the isotherms of the solutes;
 (vii) completion of the displacement chromatographic separation
       with the selected displacer and collection of fractions;

and
(viii) analysis of the collected fractions in the elution mode to
determine the extent of the separation, the purity of the
fractions and the yield.
    In the initial experiments reported here we did not attempt
to optimize the separation in terms of yield and production rate.
Rather, our intent was to demonstrate that displacement
chromatographic separations are feasible on a chiral stationary
phase, cyclodextrin-silica, and gather preliminary information
regarding the structure of displacers which can be used with
cyclodextrin-silicas. The method development sequence described in
the previous paragraph will be followed in the discussion of the
results.

Factors Which Influence Solute Retention on Cyclodextrin Silicas

Generally, cyclodextrin-silicas are used in the reversed-phase
mode. The most important retention controlling parameters to
consider are the type and concentration of the organic modifier,
and the pH, concentration and type of the buffer in the eluent.

Concentration of the Organic Solvent in the Eluent. As in other
reversed-phase systems, the solute retention curves (log k' vs. %
organic modifier) on cyclodextrin-silicas are quasi-linear, at
least over a limited concentration range. Such curves are shown in
Figs. 2-4 for the positional isomers of nitrophenol (Fig. 2) and
chloroaniline (Fig.3), and the enantiomers of Ibuprofen (Fig.4).
Similar retention curves were observed and published for other
positional isomers (67), cis/trans isomers (68) and enantiomers
(69).
    Unlike with octadecyl-silica-type reversed-phase materials,
the log k' vs. % organic modifier curves on cyclodextrin-silicas
tend to level off at very small k' values when the organic solvent
concentration of the eluent is high (generally in excess of 60 %
v/v). Armstrong proposed previously (36-39) that this behavior can
be explained by noting that at such high concentration levels the
organic solvent molecules saturate all of the cyclodextrin
cavities. The range of solute capacity factor changes on
cyclodextrin-silica rarely exceeds the two and a half orders of
magnitude limit as the concentration of the organic modifier is
varied between 0 and 100 %.
    The retention curves of the isomers or enantiomers of the
same compound tend to be closely parallel (Figs. 2-4) indicating
that in contradistinction to the octadecyl-silica phases,
separation selectivity for the isomers cannot be, in general,
improved much by changing the concentration of the organic
modifier. However, the retention curves of different compound
types, especially when the solutes have different polar functional
groups, are not parallel at all (67-69), a fortunate fact that can
be utilized to great advantage in displacement chromatography.

The Effects of the Eluent Buffer. Our understanding of the role
of the eluent buffer in the control of solute retention is
somewhat incomplete yet. It was reported that the effects of the
ionic strength depend very strongly on the charge type of the
solutes (positive, negative or noncharged) and the pH of the
eluent (67). At constant pH, the retention of noncharged solutes
increases slightly with the ionic strength of the eluent, and the
extent of the change is comparable to what is observed in
octadecyl-silica-based reversed-phase chromatographic systems. For
charged solutes the retention can both increase or decrease with
the ionic strength, depending on the particular combination of
eluent pH and solute charge (67).
    The effects of eluent pH upon the retention of various
solute types in pure water as eluent are shown in Fig. 5. The pH

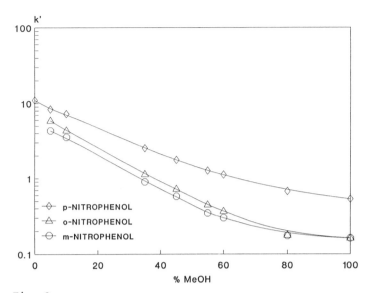

Fig. 2
Retention curves for the nitrophenol isomers on beta-cyclodextrin-silica columns. Eluent: methanol : water (% v/v), column temperature 30 C, eluent flow-rate 1 mL/min.

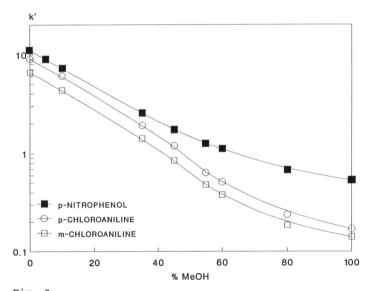

Fig. 3
Retention curves for the chloroaniline isomers on beta-cyclodextrin-silica columns. Eluent: methanol : water (% v/v), column temperature 30 C, eluent flow-rate 1 mL/min.

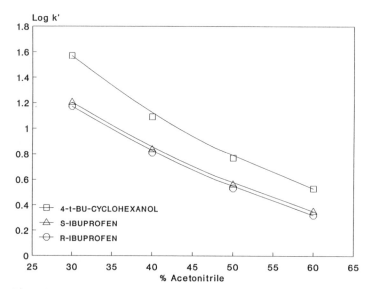

Fig. 4
Retention curves for the enantiomers of Ibuprofen on
beta-cyclodextrin-silica columns. Eluent: acetonitrile :
water (% v/v), 5 mM triethylamine adjusted to pH 6.0 by
acetic acid. Column temperature: 30 C, eluent flow-rate
1 mL/min.

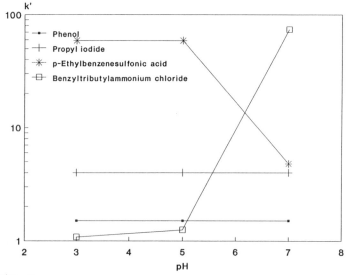

Fig. 5
Variation of solute retention as a function of the eluent
pH. Conditions: beta-cyclodextrin-silica; pure aqueous
eluent: pH adjusted by HCl. Column temperature 30 C,
eluent flow-rate 1 mL/min.

of the eluent was varied by adding hydrochloric acid, i.e. the ionic strength also varied slightly. The retention of noncharged solutes (whether nonpolar such as iodopropane or polar such as phenol) does not vary with the eluent pH. The retention of negatively charged solutes, such as the p-ethylbenzenesulfonate ion is increased significantly as the pH of the eluent is decreased from 7 to 3. When an inert salt, e.g. sodium chloride is added to the same eluent, the retention of the negatively charged solutes is decreased. In contradistinction, the retention of positively charged solutes increases very strongly as the pH of the eluent is increased (72). However, it cannot be determined yet how much of this behaviour is due to the cyclodextrin moiety and how much is caused by the silica matrix itself. More systematic studies using both chiral and nonchiral solutes will have to be completed before all the effects of the buffers upon solute retention and separation selectivity will be fully understood and utilized to enhance the separations.

The Effects of Eluent Temperature.   In agreement with previous studies (67), both solute retention and separation selectivity were found to increase significantly as the eluent temperature was reduced from 30 to 0 C, as shown in Fig. 6 for the enantiomers of tryptophan. The change is much larger than what is observed on an octadecylsilica reversed-phase column and the improved separation selectivity is attributed to more pronounced inclusion complex formation. Solute retention decreases precipitously when the temperature of the eluent is raised to or above 60 C, an important fact that can be utilized advantageously for the rapid removal of strongly adsorbed components (such as displacers) from the cyclodextrin-silica column.
Thus, in conclusion, it can be stated that retention studies such as the ones outlined above permit the selection of operation conditions (type and concentration of the organic modifier, concentration and pH of the buffer, temperature of the eluent) which lead to sufficient initial solute retention (k'>10) and maximum separation selectivity necessary for a successful displacement chromatographic separation.

For example, it can be seen in Fig. 3, that the highest possible k' values for the phenolic solutes are lower than desirable, even in pure water as eluent. The solubility of these phenolic compounds is also low in pure water. Therefore, as a compromise between the opposing requirements of sufficient retention and high solubility, a carrier solution composition of 10 % methanol : water was selected for the further studies. Similarly, it can be seen from Fig. 5, that the k' values of the Ibuprofen enantiomers are around 10 in the 30 % acetonitrile : buffer eluent, therefore this composition was selected for the carrier solution for the displacement chromatographic studies.

Individual Excess Adsorption Isotherms of Selected Displacers

Detergents have often been used as displacers on alkylsilica columns, because their solubility in aqueous systems is reasonably high and their retention can be adequately controlled by systematic variation of the hydrophylic/hydrophobic balance of the molecule. We have successfully used both ionic (67) and nonionic (68) detergents as displacers on cyclodextrin-silicas as well. Their adsorption isotherms have been published (67-69) and were found to be downwardly convex, though they did not follow the simple Langmuir adsorption isotherm equation (58). The fit to the Langmuir isotherm equation seems to be better when the displacer molecule contains at least one bulky hydrophobic group which fits the cavity of the cyclodextrin moiety well. The cationic detergent displacers often successfully reduced the untowards effects of nonspecific,

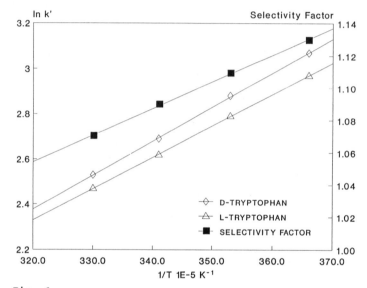

Fig. 6
Variation of solute retention and separation selectivity
as a function of the eluent temperature at constant
organic modifier concentration, pH and ionic strength.
Solutes: tryptophane enantiomers; column: alpha-
cyclodextrin-silica; eluent: 3 % v/v acetonitrile :
water, 1 mM citric acid adjusted to pH 6 by sodium
hydroxyde. Eluent flow-rate 0.5 mL/min.

ion-exchange-type solute retention as well. Both the shape of the isotherms and the amount of the adsorbed displacer strongly depended on the organic modifier concentration, pH and ionic strength of the solution (72,73).

Aromatic phenols and alcohols were also found to act as good displacers on cyclodextrin-silica columns (67,69). Since the retention studies discussed above indicate that p-nitrophenol is more retained (Fig. 2) than the chloroaniline isomers (Fig. 3), and 4-t-butylcyclohexanol is more retained than the Ibuprofen enantiomers (Fig. 4), p-nitrophenol and 4-t-butylcyclohexanol were selected as possible displacers for the separations discussed below.

Since the chloroanilines are sufficiently retained (k'>5) in a 10 % v/v methanol:water eluent, and the Ibuprofen enantiomers are sufficiently retained in a 30 % v/v acetonitrile:buffer eluent, these solvents were selected as carrier solvents for the displacement chromatographic separations. Also, these solvents were used to determine the adsorption isotherms of p-nitrophenol and 4-t-butylcyclohexanol on beta-cyclodextrin silica. The isotherms were determined from frontal chromatographic measurements as described in (56). The isotherms are shown in Figs. 7 and 8. Since both isotherms are downwardly convex, p-nitrophenol and 4-t-butylcyclohexanol might prove useful displacers for our test solutes, provided that they are more strongly adsorbed that the solutes.

Up to now, displacers had to be selected by trial-and-error. This required a significant investment of time and effort, and greatly hindered the use of cyclodextrin-silicas in the displacement chromatographic mode of operation. However, recently we recognized that the successful displacers had common structural features: they all contained a hydrophobic anchor group which closely matched the size of the cyclodextrin cavity, and a middle section which was capable of hydrogen bonding with the secondary hydroxyl groups of the cyclodextrin moiety. By connecting a third, solubility-adjusting section to the molecule, a series of structurally related displacers could be designed and synthesized. Further work is underway in our laboratory to determine the retention characteristics and adsorption isotherms of these displacer homologues on both alpha- and beta-cyclodextrin-silica stationary phases. These adsorption isotherm libraries will be published shortly (Vigh et al., J. Chromatogr. in preparation).

Individual Excess Adsorption Isotherms of Selected Solutes

Many trial-and-error experiments can be avoided during the development of a displacement chromatographic separation, when the isotherm of at least the most strongly adsorbed sample component is known. Therefore, as the next step, the adsorption isotherms of the most retained isomers of chloroaniline and Ibuprofen, the examples discussed above, were determined as shown in Figs. 8 and 9. It can be seen by comparing the isotherms of the solute and prospective displacer pairs that indeed p-nitrophenol can be used as a displacer for the separation of the chloroaniline isomers. The situation is more complicated with Ibuprofen and 4-t-butylcyclohexanol because their isotherms cross each other at 1.5 mM. This indicates that successful separations can be expected only below this concentration level. Other examples of crossing isotherms were also reported (69).

Representative Displacement Chromatographic Separations on Beta-Cyclodextrin-Silica

Once the composition of the carrier solution and the identity of the promising displacer are determined, the actual displacement chromatographic separation can be attempted. As an example for a

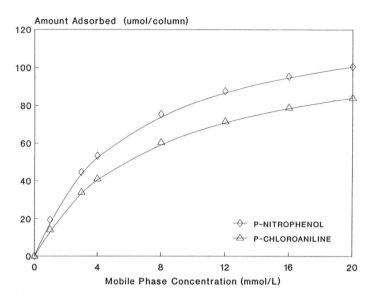

Fig. 7
Adsorption isotherms of p-nitrophenol and the most
retained isomers of chloroanilines. Conditions: 15 cm x
4.6 mm ID column packed with beta-cyclodextrin-silica;
solvent: 10 % v/v methanol : water; temperature: 30 C.

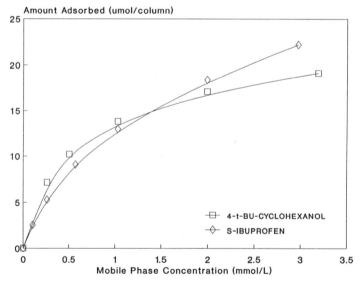

Fig. 8
Adsorption isotherms of 4-t-butylcyclohexanol and the
more retained enantiomer of Ibuprofen. Conditions: 15 cm
x 4.6 mm ID column packed with beta-cyclodextrin-silica;
solvent: 30 % v/v acetonitrile : water, 5 mM
triethylamine adjusted to pH 6.0 by acetic acid; column
temperature: 30 C.

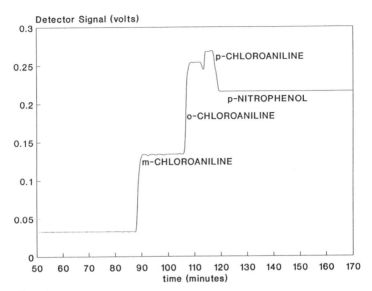

Fig. 9
Displacement chromatogram of a chloroaniline mixture containing 44.7 umol of the meta isomer, 29.8 umol of the ortho isomer and 16.4 umol of the para isomer. Conditions: 50 cm x 4.6 mm ID beta-cyclodextrin-silica column, carrier solvent: 10 % v/v methanol : water; displacer: 13 mM p-nitrophenol dissolved in the carrier solvent; displacer flow-rate: 0.2 mL/min; column temperature: 30 C.

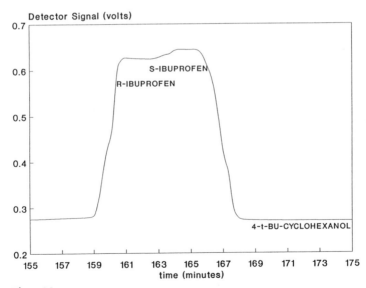

Fig. 10
Displacement chromatogram of a racemic mixture of 0.5 umol Ibuprofen. Conditions: 50 cm x 4.6 mm ID beta-cyclodextrin-silica column, carrier solvent: 30 % v/v acetonitrile : water, 5 mM triethylamine adjusted to pH 6.0 by acetic acid; displacer: 1 mM 4-t-butyl-cyclohexanol dissolved in the carrier solvent; displacer flow-rate: 0.2 mL/min; column temperature: 30 C.

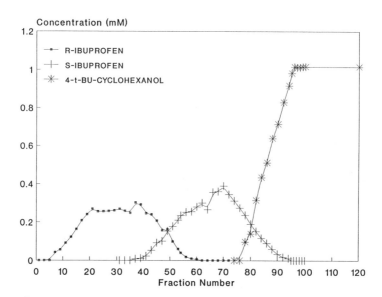

Fig. 11
Reconstructed displacement chromatogram of a racemic
mixture of 0.5 umol Ibuprofen. Conditions: as in Fig. 11.
Collected fraction size: 60 uL. The collected fractions
were analyzed on a 25 cm x 4.6 mm ID beta-cyclodextrin-
silica column using a 70 % acetonitrile : water eluent,
which contained 5 mM triethylamine and its pH was
adjusted to 6.0 by acetic acid.

positional isomer separation, the displacement chromatogram of a
0.1-millimole chloroaniline sample was obtained on a 50 cm x 4.6
mm ID. analytical beta-cyclodextrin-silica column using a 13 mM p-
nitrophenol solution as displacer. The displacement chromatogram
recorded by a UV detector is shown in Fig. 9. Analysis of the
fractions collected on the first, second and third plateaus
indicates the presence of pure meta-, ortho- and p-isomer,
respectively.

The displacement chromatogram of a 0.5-umol sample of racemic
Ibuprofen with 1 mM 4-t-butylcyclohexanol as displacer is shown in
Fig. 10. Since partially separated bands of the enantiomers can
also lead to constant plateaus in the chromatograms, the detector
trace recorded during the displacement chromatographic separation
generally does not provide sufficient information to judge the
quality of the separation. Therefore, small, 60 uL fractions were
collected from the interesting segment of the chromatogram (between
150 and 170 minutes) and were analyzed subsequently in the elution
mode using another beta-cyclodextrin-silica column and an eluent of
70 % acetonitrile : buffer. The reconstructed displacement
chromatogram of the Ibuprofen sample is shown in Fig. 11. Due to
boundary overlaps at both the front and the rear, the yield at a
given purity level is lower for the more strongly adsorbed
enantiomer.

Similar displacement chromatographic separations have been
obtained for other samples as well including naphthol isomers (67),
nitroaniline isomers (67), nitrophenol isomers (67), the cis- and
trans-isomers of 3-hexen-1-ol (68), and the enantiomers of
mephobarbital (69), hexobarbital (69), dansyl leucine (69) and
dansyl valine (69). Sample loadings on the 4.6 mm ID. analytical
columns varied between 0.1 milligram and 60 milligram; the
concentration of the separated solutes in the collected fractions
ranged from 0.1 mM to 10 mM.

Conclusions

We have briefly demonstrated in this paper that the unique
separation selectivity of beta-cyclodextrin silicas, as known from
elution chromatographic separations, can also be successfully
utilized for preparative displacement chromatographic separations.
Retention studies of the samples permit the selection of the
composition of the carrier solvent, while those of a number of
structurally related compounds facilitate the selection of a few
displacer candidates. The adsorption isotherms of these candidates
are determined and compared with those of the most strongly
retained solute, and an actual displacer and its concentration is
selected. If necessary, the separation conditions can be modified
according to the results of the first displacement chromatogram.
Elution of the least strongly adsorbed component has been observed
when the sample contained solutes of very different molecular
structure and polarity, but not when two structurally related
isomers or enantiomers were to be separated from each other.
Currently, the lack of a large number of easily available and well-
characterized displacers hinders the wider use of displacement
chromatography in the preparative separation of isomers and
enantiomers. Further work is under way in our laboratory to
synthesize several sets of rationally designed, broadly applicable
displacers for both alpha- and beta-cyclodextrin-silica columns.

Acknowledgment

Partial financial support of this work by the Texas Coordination
Board of Higher Education TATR Program, Grant Number 3376, the
National Science Foundation, Grant Number CHE-8919151 and the

196   ANALYTICAL BIOTECHNOLOGY

Minority Access for Research Careers Program, National Institute of
Health, Grant Number 5F31GM11689 is acknowledged. The authors are
grateful to Dr. Thomas Beesley of ASTEC for the beta-cyclodextrin
silica stationary phases used in this study.

Literature Cited

1. Barrett, A. M. Br. J. Pharmacol. 1968, 34, 43
2. Kawashima, K.; Levy, A.; J. Pharmacol. Exp. Ther. 1976, 196, 517
3. O'Reily, R. A. Clin. Pharmacol. Ther. 1974, 16, 348
4. Lewis, R. J.; Trager, W. F.; J. Clin. Invest. 1974, 53, 1781
5. Rummel, W.; Bradeburger, U.; Med. Pharmacol. Exp. 1976, 16, 496
6. Kupfer, A.; Bircher, J.; J. Pharmacol. Exp. Ther. 1979, 209, 190
7. Mennini, T.; Caccia, S.; Garratini, S.; Psychopharmacology
   1985, 85, 111
8. Richards, R. P.; Caccia, S.; Jori, A.; In Bioactive Analytes;
   Reid, E., Ed.; Plenum: New York, 1986; p. 273
9. Pirkle, W. H. In Chromatography and Separation Chemistry:
   Advances and Developments; Ahuja, S., Ed.; ACS Symposium Series
   No. 297; American Chemical Society: Washington, D.C., 1986;
   p.101
10. Pirkle, W. H.; Pochapsky, T. C. In Advances in Chromatography
    Giddings, J. C., Ed.; Marcel Decker: New York, 1987; p.73
11. Bender, M. I.; M. Komiyama, M. Cyclodextrin Chemistry; Springer
    Verlag: Berlin, 1978
12. Szejtli, J. Cyclodextrins and Their Inclusion Complexes;
    Akademiai Kiado: Budapest, 1982
13. Shibata, T.; Okamoto, I.; Ishii, K.; J. Liquid Chromatogr. 1986,
    9, 313
14. Szejtli, J.; Zsadon, B.; Cserhati, T. In Ordered Media in
    Chemical Separations; Hinze, W. L.; Armstrong, D. W., Eds.; ACS
    Symposium Series No. 342; American Chemical Society: Washington,
    DC, 1987; 200
15. Armstrong, D. W.; J. Liquid Chromatogr. 1984, 7, 353
16. Armstrong, D. W.; Anal. Chem. 1987, 59, 84A
17. Spino, L. A.; Armstrong, D. W. In Ordered Media in Chemical
    Separations; Hinze, W. L.; Armstrong, D. W., Eds.; ACS
    Symposium Series No. 342; American Chemical Society: Washington,
    DC, 1987; 235
18. Yamamoto, Y.; Inoue, Y.; J. Carbohydrate Chemistry 1989, 8, 29
19. Zsadon, B.; Decsi, L.; Szilasi, M.; Tudos, F.; Szejtli, J.; J.
    Chromatogr. 1983, 270, 127
20. Fujimura, K.; Ueda, T.; Ando, T.; Anal. Chem. 1983, 55, 446
21. Kawaguchi, Y.; Tanaka, M.; Nakae, M.; Funazo, K.; Shono, T.;
    Anal. Chem. 1983, 55 1852
22. Armstrong, D. W. U.S. Patent 4 539 399, 1984
23. Armstrong, D. W.; DeMond, W.; J. Chromatogr. Sci. 1984, 22, 411
24. Armstrong, D. W.; Alak, A.; DeMond, W.; Hinze, W. L.; Riehl, T.
    E.; J. Liquid Chromatogr. 1985, 8, 261
25. Cyclobond Handbook, Astec, Inc.: Whippany, NJ, 1988
26. Connors, K. A.; Pendergast, D. D.; J. Am. Chem. Soc. 1984, 106,
    607
27. Armstrong, D. W.; DeMonde, W.; Alak, A.; Hinze, W. L.; Riehl,
    T. E.; Bui, K. H.; Anal. Chem. 1985, 57, 234
28. Chang, C. A.; Abdel-Aziz, H.; Melchor, N.; Wu, Q.; Pannell, K.
    H.; Armstrong, D. W.; J. Chromatogr. 347 (1985) 51
29. Chang, C. A.; Wu, Q.; Armstrong, D. W.; J. Chromatogr. 1986,
    354, 454
30. Chang, C. A.; Wu, Q.; Tan, L.; J. Chromatogr. 1986, 361, 199
31. Chang, C. A.; Wu, Q.; J. Liquid Chromatogr. 1987, 10, 1359
32. Snider, B. G.; J. Chromatogr. 1986, 351, 548
33. Issaq, H. J.; McConnell, J. H.; Weiss, D. E.; Williams, D. G.;
    Saavedra, J. E.; J. Liquid Chromatogr. 1986, 9, 1783
34. Issaq, H. J.; Glennon, D. T.; Weiss, D. E.; Chmurny, G. N.;
    Saavedra, J. E.; J. Liquid Chromatogr. 1986, 9, 2763
35. Tindall J. W.; J. Liquid Chromatogr. 1987, 10, 1077

36. Armstrong, R. D.; Ward, T. J.; Pattabiraman, N.; Benz, C.; Armstrong, D. W.; J. Chromatogr. 1987, 414, 192
37. Hinze, W. L.; Riehl, T. E.; Armstrong, D. W.; DeMond, W.; Alak, A.; Ward, T. J.; Anal. Chem. 1985, 57, 237
38. Armstrong, D. W.; DeMond, W.; Czech, B. P.; Anal. Chem. 1985, 57, 481
39. Armstrong, D.W.; Ward, T. J.; Czech, A.; Czech, B. P.; Bartsch, R. A.; J. Org. Chem. 1987, 150, 5556
40. McClanahan, J. S.; Maguire, J. H.; J. Chromatogr. 1986, 381, 438
41. Florance, J.; Galdes, A.; Konteatis, Z.; Kosarych, Z.; Langer, K.; Martucci, C.; J. Chromatogr. 1987, 414, 313
42. Maguire, J. H.; J. Chromatogr. 1987, 387, 453
43. Ward, T. J.; Armstrong, D. W.; J. Liquid Chromatogr. 1986, 9, 407
44. Dappen, R.; Arm, H.; Meyer, V. R.; J. Chromatogr. 1986, 373, 1
45. Tanaka, M.; Ikeda, H.; Shono, T.; J. Chromatogr. 1987, 398, 165
46. Issaq, H. J.; J. Liquid Chromatogr. 1986, 9, 229
47. Issaq, H. J.; Weiss, D.; Ridlon, C.; Fox, S. D.; Muschik, G. M.; J. Liquid Chromatogr. 1986, 9, 1791
48. Abidi, S.L.; J. Chromatogr. 1986, 362, 33
49. Armstrong, D. W.; Yang, X.; Han, S. M.; Menges, R. A.; Anal. Chem. 1987, 59, 2594
50. Armstrong, R. D. In Ordered Media in Chemical Separations; Hinze, W. L.; Armstrong, D. W., Eds.; ACS Symposium Series No. 342; American Chemical Society: Washington, DC, 1987; p. 273
51. Hellferich, F. G.; Klein, G. Multicomponent Chromatography - Theory of Interference; Marcel Decker: New York, 1970
52. Kalasz, H.; Horvath, Cs.; J. Chromatogr. 1981, 215, 295
53. Horvath, Cs.; Nahum, A.; Frenz, J. H.; J. Chromatogr. 1981, 218, 365
54. Horvath, Cs.; Frenz, J. H.; El Rassi, Z.; J. Chromatogr. 1983, 255, 273
55. Horvath, Cs.; Melander, W. R. In Chromatography; Heftmann, E., Ed.; Elsevier: Amsterdam, 1983; p. 28
56. Jacobson, J.; Frenz, J. H.; Horvath, Cs.; J. Chromatogr. 1984, 316, 53
57. Frenz, J. H.; Horvath, Cs.; AIChE Journal 1985, 31, 400
58. Horvath, Cs. In The Science of Chromatography; Bruner F., Ed.; J. Chromatogr. Library, Vol. 32; Elsevier: Amsterdam, 1985; p. 179
59. Vigh, Gy.; Varga-Puchony, Z.; Szepesi, G.; Gazdag, M.; J. Chromatogr. 1986, 386, 353
60. Guiochon, G.; Katti, A.; Chromatographia 1987, 24, 165
61. Jacobson, J.; Frenz, J. H.; Horvath, Cs.; Ind. Eng. Chem. Res. 1987, 26, 43
62. Cramer, S. M.; El Rassi, Z.; LeMaster, D. M.; Horvath, Cs.; Chromatographia 1987, 24, 881
63. Cramer, S. M.; El Rassi, Z.; Horvath, Cs.; J. Chromatogr. 1987, 394, 305
64. Cramer, S. M.; Horvath, Cs.; Prep. Chromatogr. 1988, 1, 29
65. Subramanian, G.; Phillips, M. W.; Cramer, S. M.; J. Chromatogr. 1988, 439, 341
66. Phillips, M. W.; Subramanian, G.; Cramer, S. M.; J. Chromatogr. 1988, 454, 1
67. Vigh, G.; Quintero, G.; Farkas, G.; J. Chromatogr. 1989, 484, 237
68. Vigh, G.; Farkas, G., Quintero, G.; J. Chromatogr. 1989, 484, 251
69. Vigh, Gy.; Quintero, G.; Farkas, Gy.; J. Chromatogr. 1989, CHROMSYMP 1677
70. Bartha, A.; Vigh, Gy.; J. Chromatogr. 1984, 285, 44
71. Armstrong, D. W. In Ordered Media in Chemical Separations; Hinze, W. L.; Armstrong, D. W., Eds.; ACS Symposium Series No. 342; American Chemical Society: Washington, DC, 1987;
72. in preparation, Anal. Chem. 1990
73. Bartha, A.; Vigh, Gy.; Billiet, H.; de Galan, L.; J. Chromatogr., 1984, 303, 29

RECEIVED April 26, 1990

INDEXES

Author Index

Affiliation Index

Subject Index

Other ACS Books

Chemical Structure Software for Personal Computers
Edited by Daniel E. Meyer, Wendy A. Warr, and Richard A. Love
ACS Professional Reference Book; 107 pp;
clothbound, ISBN 0–8412–1538–3; paperback, ISBN 0–8412–1539–1

Personal Computers for Scientists: A Byte at a Time
By Glenn I. Ouchi
276 pp; clothbound, ISBN 0–8412–1000–4; paperback, ISBN 0–8412–1001–2

Biotechnology and Materials Science: Chemistry for the Future
Edited by Mary L. Good
160 pp; clothbound, ISBN 0–8412–1472–7; paperback, ISBN 0–8412–1473–5

Polymeric Materials: Chemistry for the Future
By Joseph Alper and Gordon L. Nelson
110 pp; clothbound, ISBN 0–8412–1622–3; paperback, ISBN 0–8412–1613–4

The Language of Biotechnology: A Dictionary of Terms
By John M. Walker and Michael Cox
ACS Professional Reference Book; 256 pp;
clothbound, ISBN 0–8412–1489–1; paperback, ISBN 0–8412–1490–5

Cancer: The Outlaw Cell, Second Edition
Edited by Richard E. LaFond
274 pp; clothbound, ISBN 0–8412–1419–0; paperback, ISBN 0–8412–1420–4

Practical Statistics for the Physical Sciences
By Larry L. Havlicek
ACS Professional Reference Book; 198 pp; clothbound; ISBN 0–8412–1453–0

The Basics of Technical Communicating
By B. Edward Cain
ACS Professional Reference Book; 198 pp;
clothbound, ISBN 0–8412–1451–4; paperback, ISBN 0–8412–1452–2

The ACS Style Guide: A Manual for Authors and Editors
Edited by Janet S. Dodd
264 pp; clothbound, ISBN 0–8412–0917–0; paperback, ISBN 0–8412–0943–X

Chemistry and Crime: From Sherlock Holmes to Today's Courtroom
Edited by Samuel M. Gerber
135 pp; clothbound, ISBN 0–8412–0784–4; paperback, ISBN 0–8412–0785–2

For further information and a free catalog of ACS books, contact:
American Chemical Society
Distribution Office, Department 225
1155 16th Street, NW, Washington, DC 20036
Telephone 800–227–5558

B.C.